TL
272.5
.W43   Weathers, Tom.
1984   Automotive computers
        and control systems

DATE			

Automotive Computers and Control Systems

Automotive Computers and Control Systems

TOM WEATHERS, JR.

CLAUD C. HUNTER

PRENTICE-HALL, INC., *Englewood Cliffs, New Jersey 07632*

Library of Congress Cataloging in Publication Data

Weathers, Tom.
  Automotive computers and control systems.

  Includes index.
  1. Automobiles—Electronic equipment. 2. Automobiles—
Motors—Control systems. 3. Automobiles—Electronic
equipment—Maintenance and repair. 4. Automobiles—Motors
—Control systems—Maintenance and repair. I. Hunter,
Claud C. II. Title
TL272.5,W43 1984   629.2′549   83-21169
ISBN 0-13-054693-3

Editorial/production supervision and interior design: Ellen Denning
Manufacturing buyer: Anthony Caruso

Printed in the United States of America

10 9 8 7 6 5 4 3 2 1

ISBN 0-13-054693-3

PRENTICE-HALL INTERNATIONAL, INC., *London*
PRENTICE-HALL OF AUSTRALIA PTY. LIMITED, *Sydney*
EDITORA PRENTICE-HALL DO BRASIL, LTDA., *Rio de Janeiro*
PRENTICE-HALL CANADA INC., *Toronto*
PRENTICE-HALL OF INDIA PRIVATE LIMITED, *New Delhi*
PRENTICE-HALL OF JAPAN, INC., *Tokyo*
PRENTICE-HALL OF SOUTHEAST ASIA PTE. LTD., *Singapore*
WHITEHALL BOOKS LIMITED, *Wellington, New Zealand*

Contents

3   POLLUTION CONTROL DEVELOPMENTS   30

4   REASONS FOR COMPUTER CONTROLS   45

5   REVIEW OF ELECTRICITY FUNDAMENTALS   49

6 SEMICONDUCTORS

7 SOLID-STATE IGNITION SYSTEMS

8 HOW COMPUTERS WORK 82

9 COMPUTER FUNCTIONS 123

13 DIAGNOSIS AND TESTING

Preface

A quiet revolution is taking place in the automotive industry. Basic engine and vehicle control functions that were once handled by mechanical devices (or by the driver) are now managed by on-board computers. Almost every car produced in the United States has some kind of computerized control.

Because the equipment used to control vehicle operation is changing, the nature of automotive repair is also changing. Manual skills must be accompanied by the ability to think in terms of logic systems and diagnostic charts. "Shade tree" guesswork will no longer get the job done; problems must be analyzed beforehand.

So, where does that leave you—someone who repairs cars or is interested in how they work? Quite simply, it leaves you out on a limb unless you are familiar with two important areas of knowledge.

Obviously, you need to know something about automotive computer functions: what the computer does and how it does it. This information will help you deal with those complex shop manuals produced by the manufacturers. Also, and perhaps not so obviously, you need to be familiar with the concept of automotive control itself, in other words, with the "why" behind computer control systems.

Until the development of computerized systems, people didn't think a lot about automotive controls. Even now, the concept of control is likely to be secondary in most people's minds to the actual computer hardware and software. After all, computers are exciting, revolutionary devices. Control, on the other hand, is an abstract topic, difficult to get a fix on.

However, it should be understood that on-board computers are just the latest in a long series of automotive control devices. You cannot truly appreciate computerized controls or do any independent problem solving without first understanding automotive controls in general. Understanding "why" is as important as understanding "what." This book covers both areas of knowledge.

Chapters 1 through 3 deal with the concept of control and with the control systems used before computers came on the scene. Chapter 4 then explains how the limitations of mechanical and manual control systems, coupled with fuel scarcity and pollution control, lead to the introduction of computer-controlled systems.

Chapters 5 through 7 review some fundamental facts about electricity. Such information is necessary not only to get an idea about how computers work, but also to understand the operation of the input and output devices that send information to and receive information from computers.

Chapter 8 provides a summary explanation of how computers work. This is the only chapter in the book that deals with the operation of the computer processor. Surprisingly, it is also the only chapter designated as being optional. Although the subject is interesting and challenging, understanding the internal operation of computers is not necessary for dealing with them. A general appreciation will do just fine. In fact, it is worth noting that the vast majority of computer software professionals (programmers) have very little knowledge about the physical operation of the computers they program.

Chapters 9 through 13 are what the preceding chapters have been preparing you for. They deal with specific computer control functions, for example, the ways that computers manage the operation of engines. Chapter 9 describes the control functions and explains their relationships to one another. The careful reader will note that many computer functions were once performed by the mechanical devices described earlier in the book. Chapters 10 and 11 then explain the operation of input and output devices, the elements in computer control systems that lend themselves most to local service. Chapter 12 reviews the operation and use of electrical test meters. Chapter 13, the last chapter in the book, introduces you to the diagnosis and testing of computer control systems.

The book will not make you an automotive computer expert, or even a control expert. However, it will give you the background necessary to approach these subjects with some confidence.

TOM WEATHERS, JR.

CLAUD C. HUNTER

Automotive Computers and Control Systems

1

The Concept
of Control

Understanding how computers operate is one thing. Understanding why computers came to be used in cars is another. Unless you happen to be a computer engineer, understanding the latter is almost as important as understanding the former.

The early chapters cover the why of computer development. In Chapter 1 we examine the concept of control. It is important because automotive computers are primarily control devices, managing the operation of other parts of the vehicle. Chapter 1 also defines some basic control categories, showing you that computers are just one kind of control. That is very important because most of the functions now performed by computers were once done by mechanical devices. After this groundwork is laid, Chapter 2 reviews the history and operation of precomputerized ignition and fuel control systems.

ASPECTS OF CONTROLS

Before comparing computers to other kinds of controls, it will be helpful to think briefly about the fundamental nature of control. Almost every mechanism, no matter how simple, requires some kind of control. Control devices existed long before computers were invented.

Consider a child's seesaw (Fig. 1-1). The action of the board is managed or *controlled* by the location of the pivot point. It is actually a control device. Children of equal weight achieve balanced operation by putting the pivot in the center of the

The position
of the pivot
controls
the see-saw.

Figure 1-1   Seesaw control.

board. Children whose weight is not the same must adjust the board back and forth
to find a balance point. In either case, the operation of the seesaw must be controlled.
Otherwise, it could not function in a useful manner

The concept of control is tied in with the basic design of a machine. For instance, in an internal combustion engine, the size and location of the cylinders controls the size and operation of the pistons. In a house design, the size and location
of the window controls the flow of air through the house (Fig. 1-2).

Even the way we use words points out the basic nature of control. Well-de-

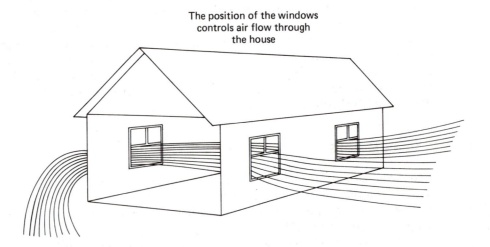

The position of the windows
controls air flow through
the house

Figure 1-2   Windows as air control.

signed, properly operating machines are said to function in a *controlled* manner. They perform their job in a predictable way. On the other hand, malfunctioning machines are said to be *uncontrolled*. They do not work the way they are supposed to and are therefore considered unpredictable and not useful.

Control Means Intelligence

Another aspect of control is the thought behind it. The control function is a deliberate attempt by a person to manage the behavior of a device. It is one of the ways in which we use to control nature. You might say that it is *our* nature to establish control over things.

Control is how we separate ordinary objects (such as sticks and stones) from useful tools. An ordinary object (such as a stick or stone) can become a tool when a person uses it in a controlled way. An even higher degree of utility is obtained when the object is reshaped for better or more precise control (Fig. 1-3).

The tool is a mirror of the man's mind.

Figure 1-3 Prehistoric man modifying a piece of flint to achieve better control.

Control As Information Processing

You have probably heard the term *information processing* applied to computers. Controlling the operation of a device is also a form of information processing (particularly if you view the terms "processing" and "information" in a general way).

Consider the child's seesaw again. The weight of the children at either end can be viewed as information. The pivot, the control point, acts as an information processor. Depending on the information input to the processor, the board will tilt up, tilt down, or stay in balance.

Most machines perform a similar sort of information processing. In the previously noted house design, the pressure and movement of air outside the house can be considered as information and the windows as information processors. The information, plus the condition of the processors (open or shut), controls the flow of air inside the house.

In an engine, the shape and location of the pistons is information and the block is the information processor (Fig. 1-4). Although the block does not actually control the operation of the pistons, it does determine their size and location. So, in effect, the block processes the piston design.

The characteristic feature of all these mechanical information processors is their inflexibility. To change the way in which information is managed, the actual hardware must be modified. In other words, the pivot point must be relocated, the windows raised or lowered, or the block design changed.

Computers, which are more commonly thought of as information processors, are flexible. A given computer can process information, and thereby exercise control, in a variety of ways. An automotive computer, for instance, can handle information about engine speed, temperature (both of the air and engine), oxygen content, and more, to control the ignition timing and the air/fuel mixture.

What Does All This Mean?

You might be wondering how the preceding discussion relates to the task at hand, understanding computers. Simply this: the computers now used on cars are just a beginning. Virtually any operation that requires information management or control can be computerized. As economic conditions make it possible, more and more operations will shift from mechanical control to electronic control. To keep up with what is happening, it helps to see the larger picture. General Motors (GM) has already provided a startling example of how computer applications can be extended beyond basic engine control.

We saw before that an engine block is actually a limited sort of information processor. Because the design is fixed, a four-cylinder engine normally acts like a four-cylinder engine, a six-cylinder like a six-cylinder, and an eight-cylinder like an eight-cylinder. Each provides a fixed level of operation. In other words, each processes information in a certain way to yield certain results.

However, what if you wanted to achieve the features of all these cylinder con-

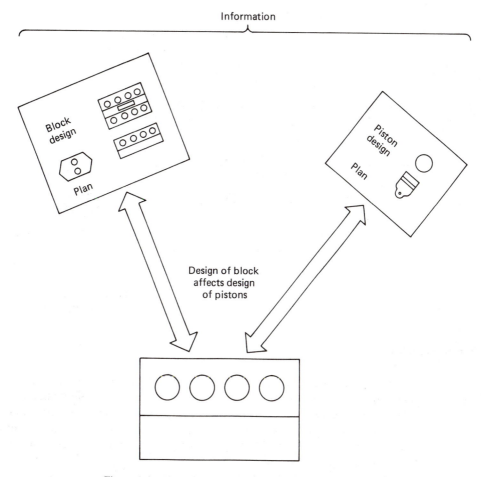

Figure 1-4 The engine block as an information processor.

figurations in one engine: eight cylinders for starting out, six for acceleration, and four for cruising? Such a task would be almost impossible for a mechanical control system. But it is possible for a flexible electronic information processor. That is what GM actually did on certain models in 1981. Using information obtained from sensors located around the engine, the computer "tells" the valve train, ignition, and fuel systems to drop out two or four cylinders when they are not needed (Fig. 1-5).

GM also uses a computer in selected models to control the torque convertor. Once certain conditions have been met, it locks the torque convertor in gear. That way, normal convertor slippage is reduced and economy improved. The computer replaces some of the information processing previously performed mechanically by the automatic transmission. In effect, it gives features of both automatic and manual transmissions.

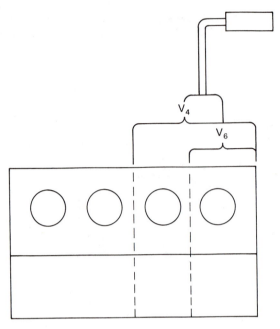

Figure 1-5 Controlling the block configuration with a computer.

THE FUTURE

Considering the applications already produced, it does not take a great deal of imagination to see how far things might go. For instance, computers could manage the operation of the suspension system, providing soft, firm, or hard control as needed. Or computers could provide antiskid braking by basing brake application on information provided by wheel travel sensors. Computers could also perform such diverse functions as controlling the windshield wipers, headlights, cruise control, and air-conditioning system. It is conceivable that a computer system could even replace the valve camshaft to control the operation of the valves directly. The list goes on and on.

LEVELS OF CONTROL

Now it is time to summarize the preceding observations by identifying some general control categories.

Static Shape Control

The most fundamental level of control comes from the very shape of a tool or machine. As we have already seen, the shape of an engine block helps control the operation of the engine. Viewed as an information processor, the block manipulates information factors external to the engine to provide new information in terms of performance.

Simple tools are other good examples of static shape control. Screwdrivers, hammers, pliers, and wrenches are controlled by their shape. Of course, the operator also has a hand in it. But even so, change the shape of any tool and the control requirements change.

These control devices also have another common factor. They are static with respect to time and space (Fig. 1-6). In other words, the control function does not depend on any moving parts and, unless wear or deterioration set in, it will remain the same forever. A screwdriver acts the same now as it does one year from now or one thousand years from now.

Dynamic Shape Control

Dynamic control refers to moving objects. To see what this means, think about the way a camshaft works. Its shape or contour certainly provides control: determining when and how fast valves open and close, controlling breaker-point operation in old-style distributors, and so on. However, you would not call a camshaft a static de-

Although the man grows old, the wrench still works the same way.

Figure 1-6 Static shape control is forever.

vice. The design or shape might be fixed, but its operation is not. In fact, a camshaft's control function takes place only when the cam is moving (Fig. 1-7). A stationary camshaft does not do anything. So a camshaft is dynamic with respect to space because it must move to provide control. It is dynamic with respect to time because no point on the cam surface stays in one place more than an instant.

Dynamic shape control is provided by any device whose configuration and movement control the operation of another device. Some examples are gear sets, chain and sprocket assemblies, and crankshafts.

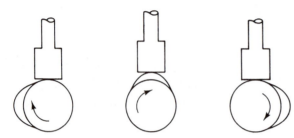

Follower goes up and down only
when cam is in motion

Figure 1-7 Dynamic shape control depends on motion.

Feedback Control

Most machines have both dynamic and static control elements. We can think of the static control components as providing a framework or skeleton for the dynamic control parts.

Many machines also need another kind of control. Take your wristwatch, for example. The configuration of the body and the design of the dial provide a certain kind of static shape control. If it is an old-style watch, the spring and timing gears provide a high degree of dynamic control. But suppose that the watch starts to run fast or slow. Then you need to be able to adjust it one way or another. This is what we call *feedback control*. It involves adjusting the operation of a machine in response to changing conditions.

In the case of the wristwatch, the feedback control is not automatic. You are an active part of the operation. Until several hundred years ago, this is how most feedback control was done. The machines were fairly simple, often driven by human or animal power. If any control adjustments were needed, the person attending the machine could make the required corrections.

However, the situation changed at the beginning of the Industrial Revolution with the invention of the steam engine. Short of constantly looking at dials and gauges, there was no way for a human operator to tell if a steam boiler was about to blow up or if a shaft was running too fast. Automatic feedback controls had to be provided in the form of pressure relief valves and governors.

Today, almost all complex machinery requires some kind of automatic feedback control. Such devices represent a different order of control and information management than the control elements mentioned previously. Feedback controls come closer to duplicating the human decision-making process. In comparison, dynamic and static shape controls are inflexible. They process information in the same way regardless of the circumstances.

Automobiles possess a variety of automatic feedback controls. One example is the thermostatic control valve in the cooling system. At a present temperature, it opens to let coolant flow from the radiator through passages in the engine block. If the engine cools too much, the valve closes again. The thermostat is the designer's representative in the engine, responding as the designer would to the information presented (Fig. 1-8).

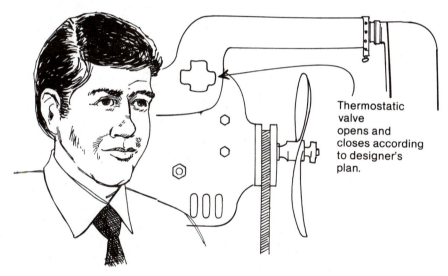

Thermostatic valve opens and closes according to designer's plan.

Figure 1-8 Automatic feedback control represents the designer's intentions.

Another "representative" is the starter drive engagement mechanism. After the engine reaches a certain speed, the drive pinion is moved out of mesh with the flywheel, thus preventing damage to the starter motor. Other examples include the automatic advance mechanism in old-style distributors, carburetor control circuits, voltage regulators, and so on.

Computerized Controls

Computers are also used as automatic feedback control units. In automotive applications, they perform many of the same functions as those handled by mechanical devices. In most cases, you would have trouble telling the difference between results obtained by computer controls versus those obtained mechanically. The real difference lies in the variety of results available. Computers provide control over a wider range of circumstances. In other words, computers can manage more information.

Why? Unlike the mechanical devices previously described, computer operation does not depend on the presence of relatively large physical shapes. Instead, the designer's intention or program is stored as a pattern of very small electrical circuits (Fig. 1-9). Simply speaking, it is possible to pack more control information into a much smaller space (Fig. 1-10).

Another difference involves versatility. Although computers, like mechanical controls, operate according to fixed programs, the computer program can easily be changed by altering the circuit pattern. Using the same basic on-board computer, a manufacturer can write different programs for different engines and operating conditions. It is much easier and less expensive to reprogram a computer than it is to produce a new mechanical control device. A time may come when a certain amount of reprogramming is done at the local shop level—which makes learning about computers even more important.

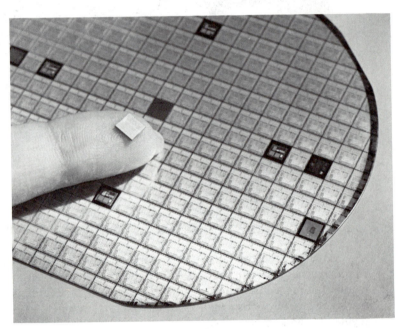

Figure 1-9 A tremendous amount of control information is packed into tiny circuits on this silicon chip. (Courtesy of Delco Electronics Division, General Motors Corporation.)

Figure 1-10   The actual computer is not much bigger than a paperback book. (Courtesy of Delco Electronics Division, General Motors Corporation.)

2

Fuel and Ignition Control Systems

When all is said and done, the basic job of an internal combustion engine is to deliver the right proportions of air and fuel inside the combustion chamber, then to ignite the mixture at the right time. If these operations are not controlled properly, nothing else makes much difference because nothing else can happen.

Chapter 2 examines fuel and ignition control systems. The basic control requirements are identified and some of the hardware used in early automobiles is reviewed. After that, in Chapter 3, we cover pollution controls, which became necessary when ignition and fuel system controls were judged insufficient to do the job alone.

EARLY AUTOMOBILES

You often read or hear people say that new cars have gotten too complicated, that they are difficult to understand and fix. The point is hard to dispute. In some engine compartments you have trouble even seeing the block because it is covered by a maze of pipes, hoses, and wires. Earlier engines, in comparison, were a marvel of simplicity (Fig. 2-1).

However, there is another side to the coin. Although earlier cars might have been easier to fix, they were not easier to drive. That is because most of the machinery in these old cars was devoted to static or dynamic shape control (explained in Chapter 1). Very little automatic feedback was provided. The driver had to make many of

12

Figure 2-1 Underhood simplicity of older car.

the periodic adjustments necessary to keep the ignition and fuel systems in the proper operating range. The ideas of the vehicle designer might be represented in the shape of the block, or in the configuration of the camshaft. But it was up to the driver to decide exactly what the ignition timing or fuel mixture ought to be. In a very real sense, the driver was an active participant in the operation of the engine.

Modern engines do not require the same degree of driver involvement. All you have to do is turn on the ignition key. Automatic feedback controls take over from there. This is one of the primary reasons for the underhood complexity of new cars (Fig. 2-2). The pipes, hoses, wires, and associated ''black boxes'' are devoted almost exclusively to sensing changing conditions and making corrective adjustments.

Some of these controls, as we will see in the discussion of fuel and ignition systems, were designed to reduce the need for driver intervention. Others, as Chapter 3 explains, were designed to automate functions that never did belong to the driver.

IGNITION SYSTEMS

Ignition controls are generally located in the distributor, the ''brain'' of the system (Fig. 2-3).

Figure 2-2   Complex engine systems of late-model car.

Basic Control Requirements

The main objective of these control sytems is to determine the exact moment when
the spark plug fires. The spark must occur at precise intervals to achieve the most
power, the best fuel economy, and the least amount of pollution. Two aspects of
timing are controlled: basic timing and timing advance.

Basic Timing

The basic timing setting refers (in points-type ignition systems) to the position of the
distributor plate and the breaker cam when the engine is at idle speeds. In newer
engines, basic timing is adjusted by loosening the distributor lock-down bolt(s) and
moving the distributor housing in a clockwise or counterclockwise direction (Fig. 2-4).
The adjustment is checked by shining a timing light on markings inscribed on one
of the crankshaft pulleys or on a plate next to the pulley.

  In early engines, timing lights were not available. Instead, a rod was placed
inside the center-located spark plug hole. As the engine was hand-cranked, the movement of the rod (and hence the piston) was observed (Fig. 2-5). When the rod reached
maximum height, the distributor was rotated until the points just opened. Such procedures are still used on many small gasoline engines. The result is a spark setting in
the neighborhood of 0 degrees TDC (Top Dead Center).

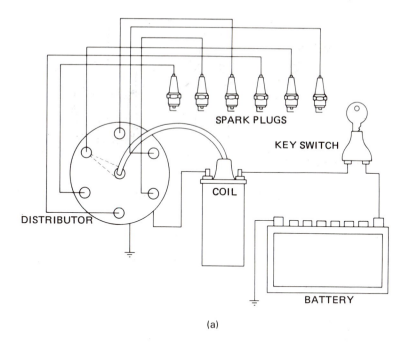

(a)

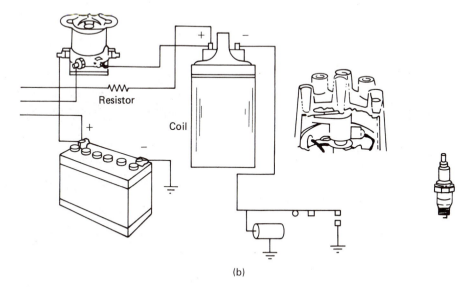

(b)

Figure 2-3 Basic ignition system: (a) main components; (b) primary current flow; (c) secondary current flow. (From T. Weathers and C. Hunter, *Fundamentals of Electricity and Automotive Electrical Systems,* Prentice-Hall, Inc., Englewood Cliffs, N.J., 1981.)

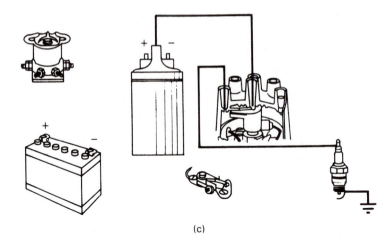

(c)

Figure 2-3 (*continued*)

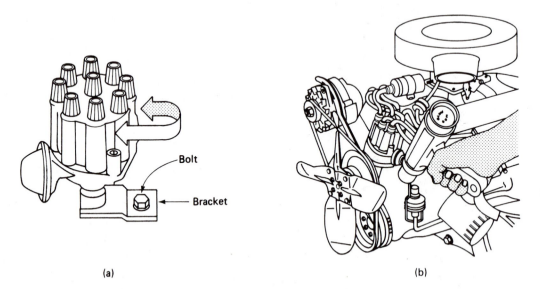

(a)   (b)

Figure 2-4 (a) Moving distributor to change timing; (b) aiming the timing light. (From T. Weathers and C. Hunter, *Fundamentals of Electricity and Automotive Electrical Systems,* Prentice-Hall, Inc., Englewood Cliffs, N.J., 1981.)

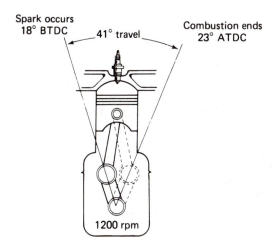

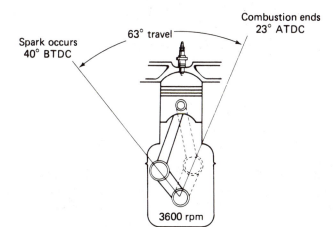

Figure 2-5 Spark timing. (From T. Weathers and C. Hunter, *Fundamentals of Electricity and Automotive Electrical Systems,* Prentice-Hall, Inc., Englewood Cliffs, N.J., 1981.)

Once set, basic timing is left unchanged. Except for the initial information provided by the timing light or control rod, no feedback is necessary. However, that is not the case with timing advance; it must change in response to changing conditions.

Timing Advance

There are two main reasons for advancing the timing (that is, for interrupting the flow of current to the coil's primary windings so that a high-voltage burst of current can be induced in the secondary winding and then routed to the spark plug). First, as the engine speeds up, there is simply less time available to burn a given charge of fuel. Second, under load conditions, the air/fuel mixture becomes upset due to the decrease in manifold vacuum for a given throttle setting. During part throttle cruising the leaner mixture does not burn well, which, again, requires an advance in spark timing.

Manual advance.   In earlier engines, timing was adjusted manually by a dash-mounted lever and rod assembly connected to the distributor breaker plate (Fig. 2-6). Moving the plate back and forth caused the points to open earlier or later. Gauging the position of the lever against a quadrant scale mounted alongside the steering wheel, the driver would retard the spark for cranking, then advance it for running. Afterward, the spark was again adjusted as engine speed and load changed.

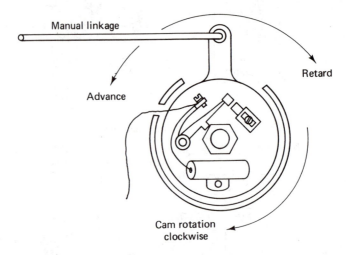

Figure 2-6   Manual timing advance.

The success of this operation depended almost entirely on the driver's ear and mechanical sense. Minor miscalculations might cause reduced economy and a loss of power. Major mistakes, particularly when attempting to turn over a hand-cranked engine, could result in kickback, which, in turn, could cause a broken arm. Therefore, in view of the hazards, and shortcomings of such a "biomechanical" feedback system, the human element was replaced by automatic feedback controls.

Centrifugal advance.   One of the first automatic feedbacks was the centrifugal advance mechanism. Located inside the distributor, it contains spring-loaded weights attached to the breaker shaft. The shaft itself is provided with a certain amount of rotational play. As the engine speed increases, the weights swing out, causing the shaft to shift position. This, in turn, causes the points to open sooner. When the engine slows down, the spring-loaded weights retract, allowing the breaker shaft to shift back in the other direction and thus retard the spark.

Such centrifugal advance mechanisms have been used up until recent times. The two principal variations have been the GM approach, which locates the weights over the breaker plate, and the approach taken by most other manufacturers, that is, locating the weights under the plate (Fig. 2-7).

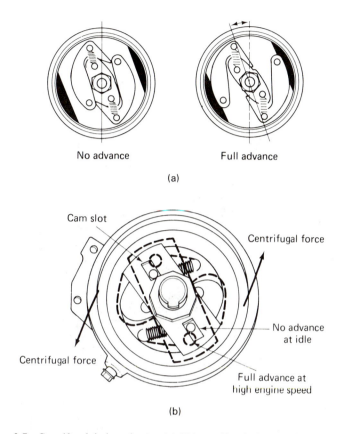

Figure 2-7 Centrifugal timing advance: (a) GM centrifugal advance; (b) centrifugal spark advance used on other engines. (From T. Weathers and C. Hunter, *Fundamentals of Electricity and Automotive Electrical Systems,* Prentice-Hall, Inc., Englewood Cliffs, N.J., 1981.)

Vacuum advance. Another sort of automatic feedback system is required to adjust the timing in response to changing engine load. At part-throttle, light-load conditions, the fuel mixture does not burn as fast, which means that the timing should be advanced. At open-throttle, heavier-load conditions, the spark should be retarded. Since manifold vacuum relates to engine load and throttle position, the feedback device most often used has been a vacuum-operated diaphragm and linkage assembly (Fig. 2-8). One side of the diaphragm is connected by a tube to a source of manifold vacuum and the other side to a linkage connected to the distributor plate. As the throttle closes and vacuum increases (pressure drops), the diaphragm causes the linkage to move the distributor plate so that the points open sooner (in points-type systems). When the throttle suddenly opens and vacuum drops (pressure goes up), a diaphragm spring causes the linkage to move the points in a retard direction.

Both the centrifugal and vacuum control systems provide infinite adjustments within certain ranges. However, interestingly enough, neither can match the adjust-

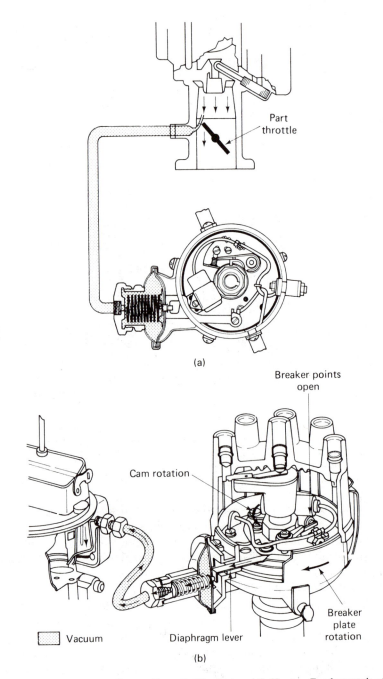

Part
throttle

(a)

Breaker points
open

Cam rotation

Breaker
plate
rotation

Vacuum   Diaphragm lever

(b)

Figure 2-8 Vacuum advance. (From T. Weathers and C. Hunter, *Fundamentals of Electricity and Automotive Electrical Systems,* Prentice-Hall, Inc., Englewood Cliffs, N.J., 1981.)

ments possible using a human opeator. A skilled driver with a keen ear can respond to more conditions than just speed and load. Unfortunately, even the most sensitive operator would have trouble responding to all the factors affecting a modern, pollution control engine. That is why computerized control systems are necessary.

FUEL SYSTEM CONTROLS

The ideal fuel mixture contains about 14.7 parts of air for every part of gasoline. However, for this *stoichometric ratio* to work properly, the correct proportions must exist throughout the mixture, almost down to the molecular level. If the mixing process is disturbed, poor performance and economy can result, even if the correct amounts of fuel and air are present.

All fuel systems contain a variety of feedback mechanisms for adjusting the fuel mixture in response to changing conditions (Fig. 2-9). Modern carburetors have special feedback circuits for starting, idle operation, high- and low-speed running, and so on. Feedback devices are also provided to warm the air going into the engine if the underhood temperature is too low, as well as to vent percolating fuel vapors if the temperature is too high. In contrast, early fuel systems had only the most basic feedback controls, relying in many instances on a human operator for guidance.

Manual Choke Controls

One control that even is today operator tended on some vehicles is the carburetor choke (Fig. 2-10). Consisting of a butterfly valve located near the top of the carburetor, the choke restricts intake air during cold cranking. This, in turn, subjects the entire throat of the carburetor to the low pressure of the intake system. The vacuum "pulls" fuel from almost every passage to create a very rich mixture going into the engine. Under ordinary circumstances, the mixture would be too rich. However, when the engine is cold, the fuel does not vaporize as readily. Therefore, the amount of fuel that is *totally* mixed with air is less than the fuel available to be mixed.

As in the case of the manually operated spark, the manual choke works well as long as the operator remains sensitive to the machinery. The problem is that many drivers leave the choke on too long. After the engine warms up, the fuel mixes more completely with air, to create an overly rich mixture. Not only will this reduce power and economy, but the raw, unburned fuel will wash oil off the cylinder walls and thereby shorten engine life.

Automatic Chokes

Automatic chokes are similar to manual chokes, except that the human operator is replaced by an independent feedback device. In most systems, whether old or new, the primary control comes from a temperature-sensitive bimetal spring. It consists of two different kinds of metals sandwiched together in a leaf or coil configuration. Like all materials, the metals expand or contract as the temperature changes. How-

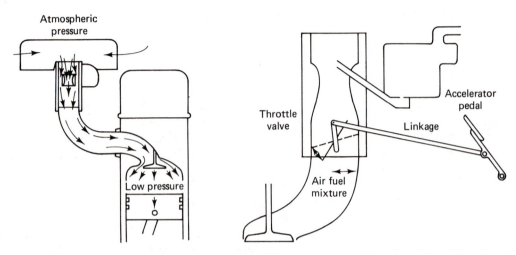

(1) Low pressure in cylinder "pulls" air through intake system (in other words, higher atmospheric pressure "pushes" air to lower pressure region)

(2) Accelerator pedal controls position of throttle

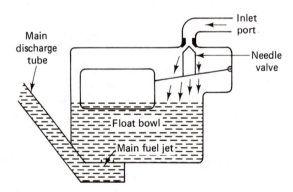

(3) Float controls needle valve and fuel delivery into bowl

Figure 2-9 Fuel system review.

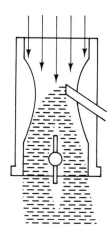

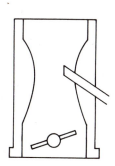

(4) When throttle is open, air flows through venturi (and manifold vacuum is low)

(5) When throttle closes, airflow is greatly reduced (and manifold vacuum is high)

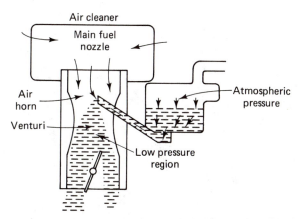

(6) Air flowing through venturi creates low pressure region which "pulls" fuel from float bowl

Figure 2-9 (*continued*)

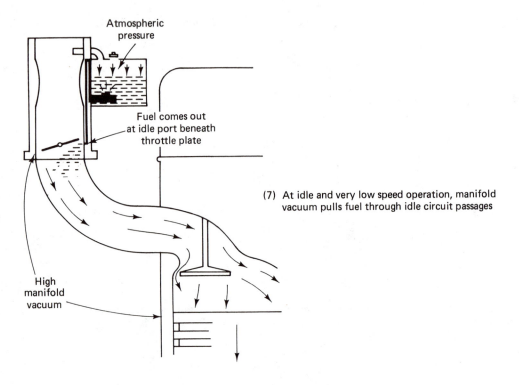

Atmospheric
pressure

Fuel comes out
at idle port beneath
throttle plate

(7) At idle and very low speed operation, manifold
vacuum pulls fuel through idle circuit passages

High
manifold
vacuum

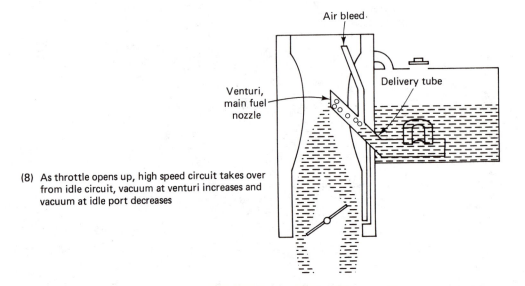

Air bleed

Delivery tube

Venturi,
main fuel
nozzle

(8) As throttle opens up, high speed circuit takes over
from idle circuit, vacuum at venturi increases and
vacuum at idle port decreases

Figure 2-9 (*continued*)

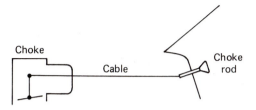

If choke is left closed too long, it will foul the
plugs, carbonize the valves, and allow liquid
gasoline to run down into the engine, diluting
the oil on the cylinder walls and
in the crankcase

Figure 2-10   Manual choke.

ever, because they are made up of different molecules, the metals expand and contract at different rates. Consequently, a bimetal leaf will bend in response to temperature changes and a bimetal coil will curl up or unwind. These movements are used to control the choke.

One early automatic choke combined a bimetal leaf spring with a small electric solenoid (Fig. 2-11). The bimetal spring was mounted on the exhaust manifold and the solenoid was attached to the choke plate by an arm and rod assembly. When the cranking motor switch was closed, the solenoid was activated and the choke closed. After the engine warmed up, the bimetal spring would bend enough to break the solenoid circuit, which let the choke open.

Another, more modern type of automatic choke uses a bimetal coil. It is mounted in an adjustable housing at the end of the choke shaft (Fig. 2-12). Air from the exhaust manifold comes up from a heat stove connected to a passage in the carburetor housing. As the air warms up, the coil unwinds and in so doing, opens the choke. In most modern systems, the coil also controls the position of a fast-idle cam. It holds the throttle open a little wider when the engine is cold, then, as the engine warms up, allows the throttle stop to return to the normal idle position.

Fuel Metering Controls

After the engine is started, new sets of conditions affect the fuel mixing process. For instance, sudden acceleration reduces manifold vacuum, causing the mixture to become lean. During high-speed operation, the air moves faster than the fuel through the intake system, which tends to upset the mixture. In addition, heavy loading alters the manifold pressure for given throttle setting, which also disrupts the balance of fuel and air.

For optimum operation, it is necessary to make compensating adjustments in the amount of fuel delivered. One early manual control system addressed the problem by running a connection from the driver's compartment to the carburetor needle valve assembly. By adjusting the position of a dash-mounted knob, the driver could vary the position of the needle with respect to its seat. This controlled the amount of fuel delivered and altered the mixture, making it richer or leaner.

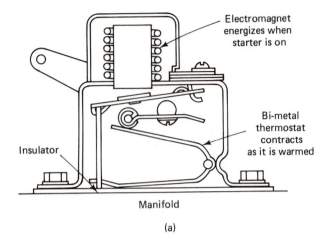

Electromagnet
energizes when
starter is on

Bi-metal
thermostat
contracts
as it is warmed

Insulator

Manifold

(a)

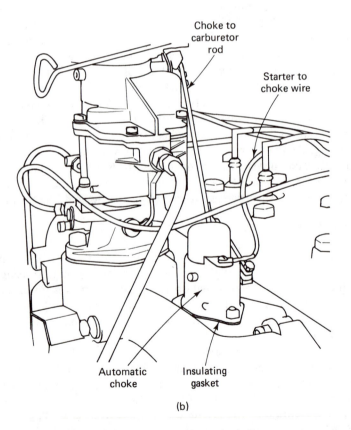

Choke to
carburetor
rod

Starter to
choke wire

Automatic
choke

Insulating
gasket

(b)

Figure 2-11 Early automatic choke.

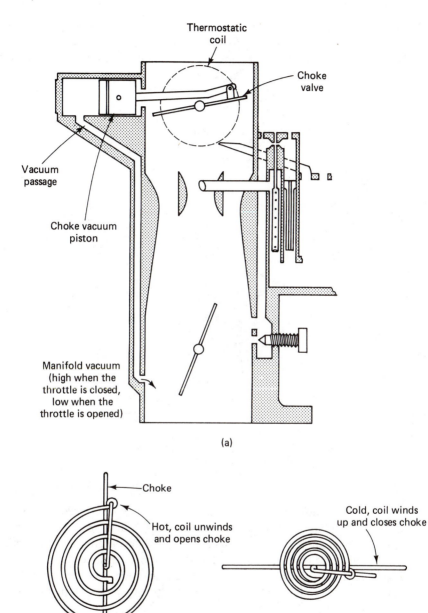

(a)

(b)

Figure 2-12   Later-model automatic choke.

Such a system was actually used on Model T Fords. Model A's, the next cars in the series, refined the process by incorporating the choke and mixture control in one knob (Fig. 2-13). The knob was pulled out for choking and twisted to adjust the mixture. Again, the problem with such systems is the lack of driver skill, or in some cases, simply the inability to a human being to react fast enough to changing conditions.

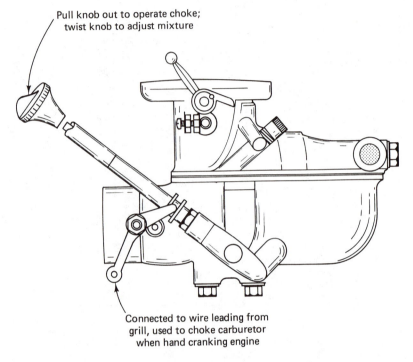

Pull knob out to operate choke;
twist knob to adjust mixture

Connected to wire leading from
grill, used to choke carburetor
when hand cranking engine

Figure 2-13   Model A Ford mixture control.

Newer carburetors provide a complex arrangement of feedback systems for adjusting fuel mixture. Extra gasoline for sustained high-speed operation is pulled by manifold vacuum through special circuits in the carburetor. The demands of sudden acceleration and load are handled by adjustable metering rods or power valves (Fig. 2-14). Operated by manifold vacuum or by linkages connected to the throttle, these systems supply additional fuel in response to the appropriate changes.

MISCELLANEOUS CONTROLS

In addition to the basic ignition and fuel system controls described in preceding sections, automobiles have a number of other feedback systems. Some, like the ones we have seen, evolved from operator-tended controls; others provided automatic feed-

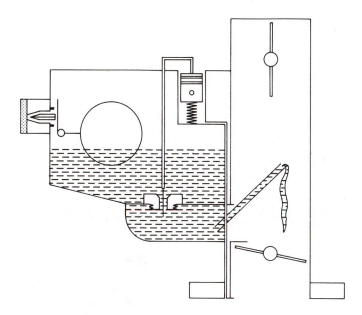

Power metering rod is held down by high vacuum at
low engine speeds. When vacuum drops . . . because
of sudden throttle opening . . . the spring pushes
the rod up which enlarges the opening and
lets more fuel flow through the jet

Figure 2-14 Power metering rod.

back from the outset. Examples include (1) starter solenoid and engagement
mechanisms, (2) automatic cruise controls, (3) automatic light dimmers, (4) automatic
windshield wiper timers, (5) self-canceling turn signals, and (6) automatic transmissions. These developments have characterized the evolution of modern automobiles,
especially American cars, where the tendency has been to isolate the driver from the
mechanical, control aspects of vehicle operation. European cars, in contrast, have
tended to emphasize the driver's role, especially in terms of suspension and steering
feedback. However, American cars are now becoming more "European" in these
respects.

3

Pollution Control Developments

The term *pollution* covers a host of twentieth-century sins. We've got beer cans in parks, chemical poisons seeping into drinking water, rivers that burn, air that chokes, and microwaves that rearrange our chromosomes. Sometimes the situation seems pretty grim. However, there is one relatively bright spot. Automakers, more than most other industries, have gone a long way toward cleaning up their products. From 1960 to 1973, they reduced unburned hydrocarbon emissions by 85% and carbon monoxide by 70%.

Responding to pressures from the federal government, the car companies have created a virtual new branch of technology. Computer controls are part of this technology. They were developed because the last vestiges of automotive pollution have been considerably more difficult to clean up than the first 70 to 80%. The feedback requirements are more exacting and comprehensive. Therefore, one of the major themes underlying this book is pollution control.

The present chapter reviews some of the antipollution measures that preceded on-board computers. We will look at the three main sources of pollution, how they have been controlled, and the combustion factors that affect pollution.

SOURCES OF POLLUTION

Automotive pollution (with the exception of NO_x) comes from gasoline or its by-products. Three main sources are the (1) crankcase, (2) fuel tank and carburetor,

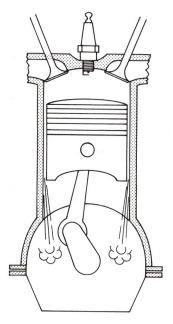

Figure 3-1 Crankcase emissions.
(Courtesy of the Chrysler Corporation.)

and (3) exhaust. The first two sources contribute 20% each or 40% of the total hydrocarbon pollution produced. The exhaust accounts for the remaining 60% (Figs. 3-1 and 3-2).

CRANKCASE EMISSIONS

Causes

Crankcase emission starts with *blow-by*. It occurs when unburned fuel and combustion by-products escape past the piston rings during the compression and power strokes. These blow-by gases fill the crankcase with contaminants. If allowed to remain, they can turn lubricating oil into sludge.

Before pollution controls, the principal means for getting rid of crankcase vapors was the road draft tube (Fig. 3-3). It was simply a pipe that vented the crankcase to the atmosphere beneath the vehicle. Air moving past the outlet created a low-pressure region which helped force vapors out of the crankcase.

Control

Dumping raw contaminants out of the road tube was never very appealing. Some readers may remember the days when black streaks were deposited between the wheel paths on all major roads. So one of the first targets of antipollution legislation was crankcase emission.

In 1963, all U.S. manufacturers installed positive crankcase ventilation (PCV) valves on their cars (Fig. 3-4). These systems operate in the following manner:

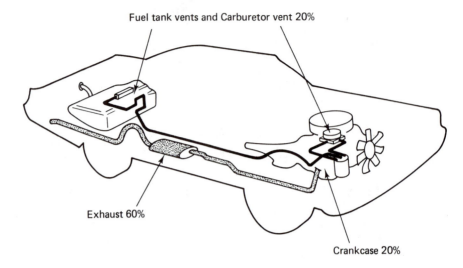

Figure 3-2   Fuel tank, carburetor and exhaust emissions.

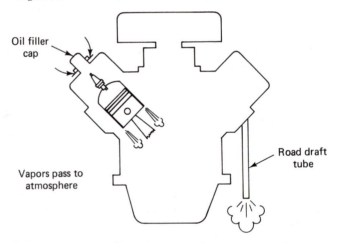

Figure 3-3   Road draft tube. (Courtesy of the Chrysler Corporation.)

1. During periods of high intake manifold vacuum, the PCV valve opens.
2. Responding to the reduced pressure, fresh air is pulled from an intake at the oil filler cap. Then, the air goes:

   1. Through the crankcase where it is mixed with contaminants. . .
   2. Through the open PCV valve. . .
   3. Into the intake manifold and then to the combustion chamber.

After 1968, the "open" PCV valves were replaced by closed systems (Fig. 3-5). They are designed to eliminate vapors pushed out of the oil filler cap during periods

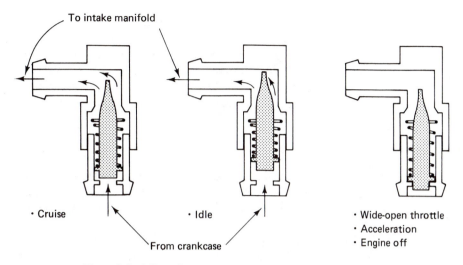

• Cruise   • Idle   • Wide-open throttle
                                                      • Acceleration
                                                      • Engine off

Figure 3-4 PCV Valve. (Courtesy of the Chrysler Corporation.)

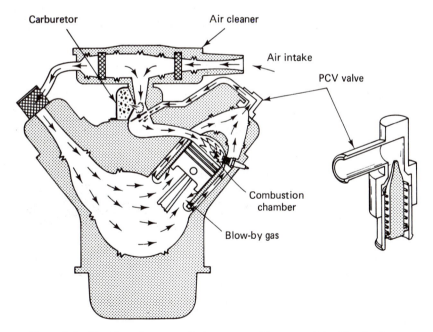

Figure 3-5 Closed crankcase ventilation system. (Courtesy of the Chrysler Corporation.)

of heavy acceleration. Instead of drawing fresh air directly from the filler cap, a supply tube goes from the cap to the air cleaner. That way, excess vapors coming from the cap are vented into the carburetor. At other times, the system works the same as before (except that fresh air is always drawn through the air cleaner). Closed PCV systems are still used. However, since they are automatic feedback devices, responding to changing conditions, their operation is now subject to computer control.

EVAPORATIVE EMISSIONS

Gasoline is a very volatile liquid. Place an uncovered container in the open and before long the entire contents will evaporate. The resulting vapor consists mostly of hydrocarbons, a pollutant. In an automobile, the two main sources of evaporative pollution are the fuel tank and carburetor float bowl.

One obvious answer for this kind of pollution is to seal off the sources of evaporation (Fig. 3-6). That is exactly what the manufacturers have done. Gas tank caps no longer allow vapors to escape into the atmosphere, and float bowls are vented into the carburetor intake or into a special storage canister.

However, just sealing the sources of evaporation is not enough. Various kinds of vents and pressure relief valves must be provided. That is because fuel flows from the tank to the pump in response to pressure differentials between the pump and the air in the tank. If the tank is completely sealed off, a vacuum will build up as the fuel is withdrawn. Such a situation would be like trying to pour liquid from a can with only one small hole. Pretty soon, the pressure inside would be less than

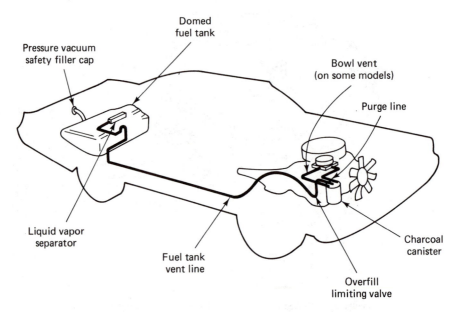

Figure 3-6   Controlling evaporative emissions. (Courtesy of the Chrysler Corporation.)

the pressure outside and the flow would stop. Therefore, gas caps must (and do) perform two functions. They seal closed when the pressure inside the tank increases and let air enter when the pressure inside the tank is reduced.

Another problem is the vapor that builds up inside closed spaces. Unless controlled, dangerous pressures and concentrations of explosive gases can result. One solution adopted by most manufacturers is a charcoal canister connected to the source of evaporation. Fuel vapors are trapped in the canister when the engine is not running. Then, when the engine is started, a fresh air purge is pulled through the canister. The air mixes with the evaporated fuel, carrying it into the engine to be burned.

Manufacturers employ a number of other devices and techniques to help control evaporative emissions. However, it is the canister-purging operation that is most likely to come under computer control. Sensitive, late-model engines cannot tolerate unplanned additions of raw, unburned vapors into the air/fuel mixture.

EXHAUST EMISSIONS

With 60% of the total, the exhaust is responsible for most automotive pollution. If you think about it, this is natural since a majority of the combustion by-products end up in the exhaust system. Consequently, considerable time and money have been spent cleaning up the exhaust. However, before looking at particular devices and techniques, it will be helpful to review some basic facts about the combustion process and the kinds of contaminants that are produced.

REVIEW OF THE COMBUSTION PROCESS

Four ingredients are necessary for combustion: air, fuel, heat, and time. Under ideal conditions, exactly 14.7 parts of air for each part of fuel will result in complete combustion with no harmful by-products left over. As we noted in Chapter 2, this is called a stoichiometric air/fuel ratio. As we also noted, problems occur if the conditions are not ideal, which they never are in the real world in which engines must operate.

Although any number of factors can affect the combustion process, the results usually fall into a limited number of categories. One common problem is incomplete mixing of air and fuel. For complete combustion to occur, the mixture must be uniform. Each molecule of fuel must be surrounded by 14.7 molecules of air. Naturally the process must start with the correct amounts of air and fuel. However, the temperature and pressure must also stay within narrow bounds. If the temperature of the air and fuel is too low, the fuel molecules will not bounce around enough to stay in a gaseous state. They will turn back into a liquid, which means that the mixture will be too lean since not enough fuel is available in a usable form. That is why engines have chokes and heat control devices.

Variations in the pressure of the air/fuel mixture can have similar effects. Generally, the lower the pressure, the easier it is for fuel to stay in a vapor state. There is less resistance to molecular movement. However, if the pressure suddenly increases, the fuel is likely to turn back into liquid droplets. This happens when the manifold vacuum drops (or the pressure increases) under load or sudden acceleration. To compensate, extra fuel is usually added during these lean periods. As a result, the overall mixture might be richer than 14.7:1, even though the effective mixture is the same or less.

Another common problem is burning time. A given quantity of a certain fuel mixture requires a definite interval for complete combustion. Therefore, when the engine speeds up, spark timing must be advanced so that the fuel will have enough time to be burned. Timing must also be advanced if the fuel mixture changes suddenly—for example, if the manifold pressure increases.

THREE PRINCIPAL EXHAUST POLLUTANTS

The targets of all engine emission control systems are three principal pollutants: hydrocarbons, carbon monoxide, and nitrogen oxide.

Hydrocarbons (HC).   These are the basic ingredients of petroleum products. When combined with oxygen in the burning process, hydrocarbons provide the energy to operate internal combustion engines. However, the HC molecules that do not burn and escape into the atmosphere are regarded as pollution. Any condition that affects combustion can cause HC emission. These conditions include overly rich carburetor or injector settings, dirty air cleaners, misfiring spark plugs, and improper timing.

Carbon monoxide (CO).   This gas has been referred to as the quiet killer. People who commit suicide by operating their cars in closed garages are victims of carbon monoxide poisoning. The resulting corpses usually have cherry lips and rosy complexions. Carbon monoxide generally occurs when there is insufficient air in the air/fuel mixture. Any condition that reduces airflow such as a dirty air cleaner can cause CO formation.

Nitrogen oxide (NO_x).   This gas, among all automotive pollutants, is not a by-product of the fuel-burning process. It is formed anywhere that very high temperatures and pressures occur, such as the combustion chamber of a late-model automotive engine. Oxygen and nitrogen, two of the most common substances in the air, combine to form various oxides of nitrogen. As combustion temperatures and pressure go up, NO_x formation increases.

CONTROLLING EXHAUST EMISSION

A number of approaches have been employed to clean up exhaust emissions. They can be grouped into three main categories:

1. Air injection system

2. Engine modification system

3. Catalytic reactor system

At different times, most manufacturers have used a combination of all three techniques.

Air Injection Systems

This method features an air pump driven by a V-belt connected to a pulley at the end of the crankshaft (Fig. 3-7). High-pressure air is injected into the exhaust stream near the root of the exhaust manifold. The extra air helps burn away HC vapors as well as eliminating some of the CO contained in the exhaust. The system acts like an afterburner. In earlier versions, air was supplied constantly to the exhaust. Newer systems operate on a periodic basis, the exact intervals controlled, in some cases, by an on-board computer. All systems, however, contain a variety of check and/or bypass valves to protect the components from backfiring, excessive pressure, failure of critical parts, and so on.

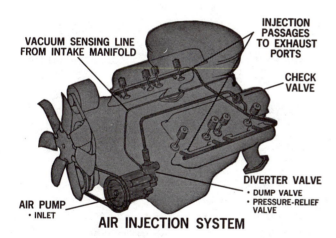

AIR INJECTION SYSTEM

Figure 3-7 Air injection system. (Courtesy of the Chrysler Corporation.)

Modified Engine System

A number of devices and techniques fall into this category (Fig. 3-8). All manufacturers have used one variation or another of the modified engine approach. Some of the trade names are:

1. ENGINE MOD, American Motors

2. IMCO (Improved Combustion System), Ford

3. CCS (Combustion Control System), GM

4. CAS (Clean Air System), Chrysler

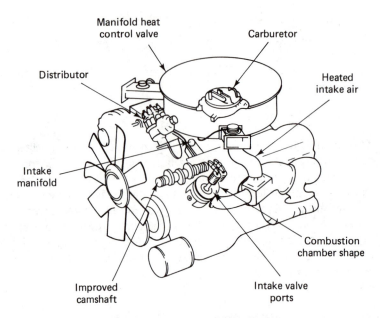

Figure 3-8 Typical engine modifications. (Courtesy of the Chrysler Corporation.)

Following is a partial list and brief description of some modified engine functions.

Reduced quench area.   *Quench* refers to the close spaces inside the combustion chamber that tend to snuff out the flame spread by burning air and fuel. Reducing the quench areas (or increasing the height of the quench zones) prolongs combustion, which reduces emission (Fig. 3-9).

Increased valve overlap.   Camshaft lobes have been redesigned so that the intake and exhaust valve opening overlaps—in other words, so that both valves are open at the transition between the exhaust stroke and the intake stroke (Fig. 3-10). This helps scavenge or wash out exhaust gases, which improves combustion and reduces emission.

Improved intake manifolds.   The air/fuel mixture can be affected by the flow characteristics inside the intake manifold. Rough surfaces will slow the movement of the mixture, which, in turn, will cause some of the fuel to condense into a liquid. Uneven passages or ports can also affect velocity and pressure. The same is true for unequal-length passages between the carburetor and the intake manifold. Since pollution control became a factor, more attention has been paid to intake manifold design.

Reduced compression ratios.   Compression ratios were reduced in the early 1970s when many engines were designed to run on lower-octane fuel. This change also resulted in lower levels of HC and NO_x.

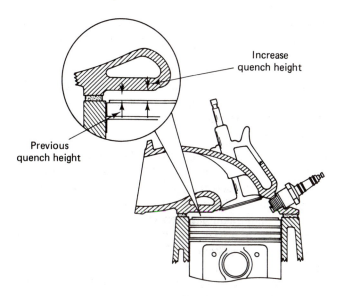

Increase
quench height

Previous
quench height

Figure 3-9 Increased quench height.
(Courtesy of the Chrysler Corporation.)

Lean calibrated carburetors.   Most prepollution engines ran on slightly rich fuel mixtures. Such carburetor settings are generally associated with improved performance and drivability. However, since rich mixtures also lead to HC and CO formation, fuel systems are now designed to operate toward the lean end of the scale. Present-day carburetors contain plugs and limiters which restrict the adjustment range to lean settings.

Improved cold-starting procedures.   Cold starting, as we have seen, is a special operation requiring a very rich fuel mixture. If not precisely controlled, it can result in incomplete burning and increased pollution. Manufacturers have provided a number of devices to warm the engine faster and to make the choke more sensitive.

Camshaft modified to alter
valve overlap

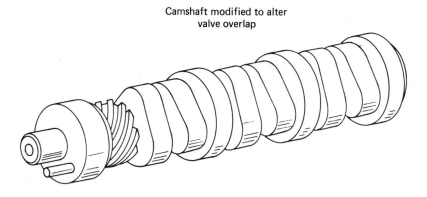

Figure 3-10 Camshaft modifications. (Courtesy of the Chrysler Corporation.)

Modified pistons. Pistons have been given special crowns and contours to help force the air and fuel into flow patterns that promote mixing and complete combustion.

Modified distributor operation. The distributor is an integral part of the combustion process since it determines the moment of ignition. Therefore, it is natural that the distributor was subjected to many pollution control-related changes. We will discuss some of those modifications next.

Duel-Diaphragm Advance. Before pollution controls, the distributor vacuum advance diaphragm responded to a source of manifold vacuum located just above the closed throttle plate. As we have noted before, it was done to improve burning during light-load, part-throttle conditions. Pollution control distributors still perform this function. However, an additional diaphragm has been added to some distributors. Connected to an outlet near the intake manifold, this diaphragm *retards* the spark when the throttle is fully closed (at idle speeds and during periods of *deceleration*). This helps reduce engine speed, which gives more time for complete combustion. Together with the retard diaphragm, manufacturers have also provided devices (such as dashpots and idle-speed solenoids) to help keep the throttle open during idle and deceleration.

Temperature-Sensitive Advance. Lean mixtures and retarded spark at idle speeds can cause an engine to run hotter. To prevent overheating, temperature-sensitive advance mechanisms have been provided (Fig. 3-11). The main elements are a three-way valve and a temperature-actuated rod positioned in or near a coolant passage.

The valve acts as a junction point between a carburetor vacuum source (above the throttle plate), a manifold vacuum source (below the throttle plate), and the distributor vacuum advance mechanism. At normal temperatures, the rod positions the

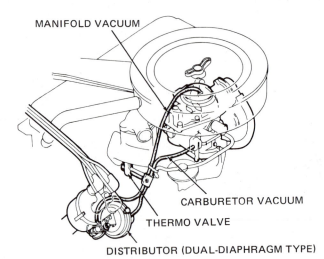

MANIFOLD VACUUM

CARBURETOR VACUUM

THERMO VALVE

DISTRIBUTOR (DUAL-DIAPHRAGM TYPE)

Figure 3-11 Dual-diaphragm and temperature-sensitive advance. (Courtesy of the Chrysler Corporation.)

valve so that the distributor responds only to carburetor vacuum. When the temperature increases above a certain level, the rod switches the valve to lower-pressure, manifold vacuum. This causes a timing advance that increases the engine speed and lowers the operating temperature.

Delayed Advance.   There is some evidence that delaying the spark advance during acceleration above certain temperatures will reduce NO_x. Chrysler has produced a device called OSAC (Orifice Spark Advance Control) to perform this function (Fig. 3-12). Located on top of the air cleaner, this valve delays advance from idle to part throttle for about 17 seconds. It operates only when the underhood temperature is above 60°F. Other manufacturers have produced units performing similar functions.

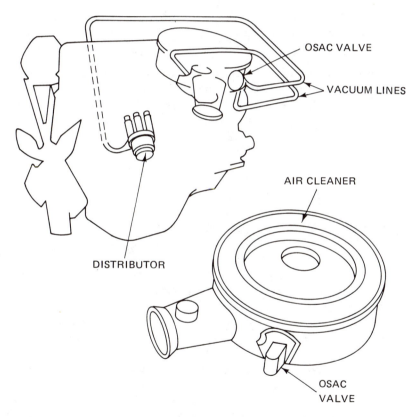

Figure 3-12   Chrysler OSAC system. (Courtesy of the Chrysler Corporation.)

Exhaust gas recirculation (EGR).   Leaving the distributor, for this chapter at least, we look next at the last engine modification on our list, the EGR system (Fig. 3-13). Introduced in 1973, it has been one of the primary methods for controlling NO_x emission. The system consists basically of small, connecting passages between the intake and exhaust manifolds and temperature- and/or pressure-sensitive valves.

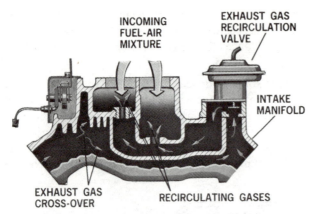

INCOMING
FUEL-AIR
MIXTURE

EXHAUST GAS
RECIRCULATION
VALVE

INTAKE
MANIFOLD

EXHAUST GAS
CROSS-OVER

RECIRCULATING GASES

EXHAUST GAS RECIRCULATION

Figure 3-13 EGR system. (Courtesy of the Chrysler Corporation.)

When the appropriate conditions are reached, the system routes a small quantity of exhaust gas to the air/fuel mixture. Since the exhaust has already been burned, it will support little if any further combustion. Diluting the air/fuel mixture with an effectively inert gas reduces the temperature of the combustion process, which, in turn, reduces the formation of NO_x.

Catalytic Reactors

As federal pollution standards became progressively tighter, manufacturers have had to find newer and more thorough ways to control exhaust emissions. One of the most widely adopted devices is the catalytic reactor or convertor. It was introduced in 1975 on all American-made cars and many imports. That was the same year in which permissible levels of HC and CO dropped respectively from 3.4 and 39.0 grams per vehicle per mile to 0.46 gram and 4.7 grams per vehicle per mile.

Located in the exhaust system, usually upstream from the muffler, a catalytic reactor looks like a small muffler or resonator (Fig. 3-14). Inside, however, the resemblance ends (Fig. 3-15). The reactor (of certain models) is filled with a honeycomb-like ceramic core. Surrounding this fragile, clay structure is steel mesh. It offers protection from road shocks and jolts.

In precomputerized engines, the core was created with a mixture of palladium and platinum. These are the active ingredients of the system, the catalytic material. When exposed to hot exhaust gases, the catalyst gets even hotter, reaching temperatures between 1300 and 1600 °F in normal operation. The catalyst itself is not changed except in the presence of very rich exhaust vapors or when leaded fuel is used. However, the catalyst promotes a reaction between HC, oxygen (O_2), and CO. As a result, the two pollutants are changed into water (H_2O) and carbon dioxide (CO_2). The latter is a harmless substance used to provide the fizz in carbonated drinks.

In 1980, a third ingredient to control NO_x was added to many catalytic reactors. We will learn more about that in Chapter 9.

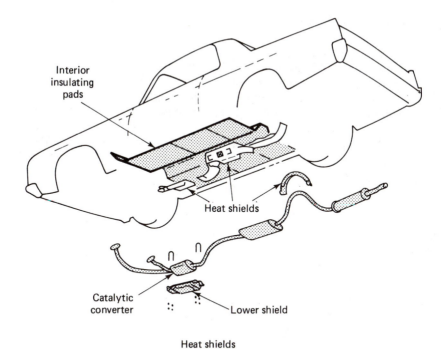

Interior insulating pads

Heat shields

Catalytic converter

Lower shield

Heat shields

Figure 3-14   Location of catalytic reactor and heat shields. (Courtesy of the Chrysler Corporation.)

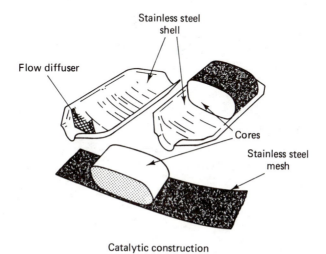

Stainless steel shell

Flow diffuser

Cores

Stainless steel mesh

Catalytic construction

Figure 3-15   Inside catalytic reactor. (Courtesy of the Chrysler Corporation.)

SUMMARY/FOREWARNING

As we have seen, many of the pollution control systems developed in the 1960s and 1970s were of the feedback variety. Most of these functions are still performed. However, as you might imagine, some of the hardware and particular approaches have changed. This is especially true for those operations that lend themselves to computer control. So when you get to the chapter on computer functions (Chapter 9), be sure to remember the requirements and processes you read about in this chapter. But do not expect the names or components necessarily to be the same.

4

Reasons for Computer Controls

The previous chapters have indicated some of the advantages of on-board computers. Compared to mechanical or electromechanical feedback systems, computers are faster, more versatile, and pack more control information into a smaller package. However, just saying that computers are better does not explain why they have been so widely adopted. To understand that, we need to look at the factors surrounding their development for use in cars.

REASONS FOR TECHNICAL CHANGE

Technical change, whether it relates to computers, TV sets, or whatever, comes about only for certain reasons. Two of the primary factors are need and profit. In the past, need and profit were two sides of the same coin. Innovations reached the market only if need was preceived to equal a widespread public desire to spend money. In other words, although necessity might be the mother of invention, profit is the father.

The innovations that shaped the first 50 or 60 years of automotive development followed this pattern. Electric self-starters, feedback devices of various kinds, automatic transmissions—all were produced for profit. These inventions may have provided great satisfaction for their creators; they may have also addressed real social needs; however, it was the public's willingness to buy that made them possible. Those that were not accepted fell by the wayside.

This pattern continued up until the middle to late 1960s. The emphasis was on power and drivability because that is how the manufacturers perceived the public's taste and buying habits. For instance, a typical American-made sedan weighed 2 tons and was powered by a 300-horsepower engine that got 11 miles to the gallon. An average "sporty car" of the period had a performance-modified version of the same engine, went from 0 to 60 miles per hour in 9 seconds, and got only 8 miles per gallon. There were occasional attempts to produce economy cars, such as the Corvair and the early Falcons and Valiants. But they did not sell well for long periods. After all, fuel was readily available and cheap. And pollution, although recognized as a problem, was not subject to strict controls.

If you view engine operation as a four-part equation whose major factors are power, drivability, pollution control, and economy, the formula was shifted in favor of the first two factors. Designers did not have to be concerned with complicated feedback control devices. Engines operated on relatively rich mixtures that were fairly simple to maintain. The basic controls that had been used for years in the carburetor and distributor remained adequate for the job (Fig. 4-1).

Figure 4-1 1966 Volvo, 1225. A prepollution control European sedan.

LEGISLATED PRIORITIES

Then, in the late 1960s and early 1970s another factor entered the picture. The federal government, responding to increases in airborne pollution, began to pass laws that competed with the public's seeming passion for big, fast cars. As a result, manufacturers embarked on a program of federally mandated change. For the first time, car companies had to consider seriously the third part of the four-part operation equation, pollution control.

However, even after emission control became a factor, the mechanical control systems remained adequate. Power and drivability did suffer slightly because of leaner mixtures and other internal and external engine changes. Yet drivability was not all that bad and it was always possible to make engines larger to compensate for the power loss. And despite the fact that larger pollution control engines of the period used more fuel, the operating costs were still within the range of most people. There was no need to make any basic changes in the control systems used to monitor and adjust air/fuel mixtures and ignition timing.

Then, in the middle to late 1970s, a number of significant events took place:

1. Fuel costs rose dramatically, because of political conditions in the Middle East and increased consumption plus inflation at home.
2. The federal government passed legislation requiring manufacturers to produce more-fuel-efficient vehicles.
3. Pollution controls became even more rigid.
4. The cost and size of computers decreased while their power and capacity increased.

So, for the first time, all four factors in the equation had to be given equal weight. Engineers had to design smaller engines for smaller cars and still provide adequate power, good drivability, improved economy, and reduced emissions.

Despite the usual changes in load, acceleration, and other operating conditions, engines had to stay even closer to the 14.7:1 stoichometric fuel mixture. As a result, conventional systems were pushed to the limit of their capacity. One solution was to rely more and more on electronic, solid-state ignition systems, such as those described in Chapter 7. These devices, which became common in the middle 1970s, were designed to produce "hotter" ignition sparking, which allowed engines to run on leaner fuel mixtures.

The problem is that these hotter, solid-state ignitions still do not provide much better control than points-type systems. Slight changes in lean-burning engines can shift the equation into the misfiring range, where power, drivability, economy, and emissions all suffer. To avoid these problems, it is necessary to take into account a number of operating factors and to be able to make almost instant changes in timing and in fuel mixture.

ERA OF ELECTRONIC CONTROLS

This is where the new generation of low-cost, high-powered computers enter the picture. Instead of using mechanical devices to sense changing conditions and make compensating adjustments, manufacturers have come to rely on electronic brains. The first was Chrysler's Lean Burn System. Introduced in 1976, it was primarily an add-on device designed to control spark timing in response to engine and air temperature, throttle position, and engine speed. Since then, all U.S. automakers have installed computer controls. Some, like the early Chrysler system, just manage the operation of the distributor. Others control a modified, but still fairly conventional carburetor. Still others control both the distributor and a new type of fuel-injection system. At least one manufacturer has provided a computer with built-in troubleshooting features (Fig. 4-2).

In the future, computers will serve more and more as the focus of a total vehi-

Figure 4-2 1983 Chevrolet Camaro.
An electronically controlled American
sporty coupe.

cle information management system. Not only will the computer control the engine,
it may also operate digital, space-saving instrument panels, electronic door locks,
automatic radio antennas, "smart" suspension and brake systems, and so on.

Part of the reason for these enhancements is the nature of the computer; it
naturally lends itself to such "systems" approaches. Another reason is economics.
Although manufacturers may have been forced by outside factors to adopt computers,
they will now use them to pursue profit. In other words, we are back where we started
at the beginning of this chapter. Public acceptance will now determine how far
computerization goes.

5

Review of Electricity
Fundamentals

The preceding chapters went over the "why" of computers and control systems. The remaining chapters cover the "how". In one way or another, all these chapters deal with electricity. Therefore, to get started out on the right foot, we review the separate but related concepts of electricity and magnetism. The first part of the chapter discusses atomic theory, current flow, various kinds of conductors, some practical effects of electrical flow, and the common terms used when measuring electricity. The next part of the chapter covers magnetic fields, various kinds of magnets, a simplified theory of how magnets work, and the concept of induced current flow. The last part of the chapter introduces some categories into which electrical control devices can be grouped and notes the chapters where these devices are discussed in more detail.

ELECTRICITY

What Is Electricity?

We often use the word *electricity* as if it described a real, physical object. However, that is not exactly what the word means. How many times have you seen, tasted, or touched electricity? Never, although you may have seen and touched the effects of electricity.

When we talk about electricity, we are actually talking about movement, a flow of tiny particles deep within the heart of matter. The visible or detectable signs of electricity—sparks, shocks, readings on test instruments, and so on—are a result of this flow. When the particles are in motion, they have an effect on the material through which they flow as well as on the surrounding space.

Nature in Balance

Why do the particles move in a wire, a computer circuit, electrical control device, or anywhere else? It has to do with the universal tendency of conflicting forces to try to stay in balance. The planets remain in their orbits because the centrifugal force of their movement is balanced by the force of gravity from the sun. Water stays in a man-made lake because the weight of the water is balanced by the strength of the dam.

All these are examples of forces in balance. However, what happens if something changes, if, for instance, the dam breaks? Then the forces are no longer in balance. In the case of the broken dam, water rushes downhill, affecting everything in its path, until the lake is emptied and a new balance is achieved.

It is this urge for balance that causes electrical flow. A battery, alternator, or some other device builds up the number of tiny electrical particles in one part of a circuit and reduces the number in another part of the circuit. If the circuit is complete—in other words, if there is an uninterrupted path between the out-of-balance parts—current flows. Excess particles flow to the part of the circuit containing fewer particles. The current continues to flow as long as the battery or alternator maintains the imbalance. Like the water flowing downhill from the broken dam, everything that lies in the path of the current is affected.

Atomic Structure

The next question is: What are these particles? If you have been exposed to the subject previously, you have probably already guessed that the particles are electrons. Electrons, together with protons, neutrons, and other particles, make up atoms, which are the basic building blocks of nature. Atoms or pieces of atoms combine in various ways to create all the material in the universe.

The simplest atomic combination is the element hydrogen, which is a very light gas at room temperature (Fig. 5-1). It has one proton at the center of the atom and one electron whirling around the proton, something like a planet whirling around the sun. Hydrogen was one of the first elements created at the beginning of the universe. Other elements, including helium, oxygen, copper, gold, and lead—more than

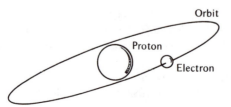

Figure 5-1 Hydrogen, the simplest element. (From T. Weathers and C. Hunter, *Diesel Engines for Automobiles and Small Trucks,* Reston Publishing Company, Inc., Reston, Va., 1981.)

100 altogether—were formed as additional protons, neutrons, electrons, and other parts were added to the original hydrogen atoms (Fig. 5-2).

The atomic parts "stick" together to create atoms because nature strives for balance everywhere, from the very small to the very large. In atoms, the balance we are interested in is between protons and electrons. Protons are said to have a *positive charge*, like the south end of a magnet. Electrons are said to have a *negative charge*, like the north end of a magnet. For an atom to be balanced electrically, it must have an equal number of electrons and protons.

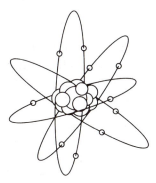

Figure 5-2 A more complex atom. (From T. Weathers and C. Hunter, *Diesel Engines for Automobiles and Small Trucks,* Reston Publishing Company, Inc., Reston, Va., 1981.)

When an element is created, protons are packed together in the center, or nucleus, of the atoms. Electrons are attracted in equal numbers to orbits surrounding the nucleus. Because the electrons have the same charge and tend to repel one another, each orbit can contain only so many electrons, two in the first orbit, eight in the second, and so on.

The last or outermost orbit is the most important in our understanding of electricity. The number of electrons in this orbit determines whether an element will be a conductor, a nonconductor, or a semiconductor of electricity. It is in this orbit that electrical flow takes place.

Electrical Flow

To see how electrical flow works, we will take an imaginary trip into the atomic structure of a copper wire. Copper, a conductor of electricity, has 29 electrons. That gives enough electrons to fill all but the last orbit, which has only one electron. Because these single electrons are so far from the nucleus and because they are not bound in a fixed pattern with other electrons, they tend to drift from atom to atom. From our imaginary vantage point, the outer orbit electrons might resemble a swarm of fireflies, wandering here and there in response to random influences (Fig. 5-3).

As long as the wire is not hooked into a live electrical circuit, the pattern of electron movement remains random. An atom in one place loses an electron. Another electron, from somewhere else, takes its place, attracted by the extra positive charge exerted by the atom that has lost its electron. Since the movements are random, the total of all the electrical charges on all the atoms remains balanced.

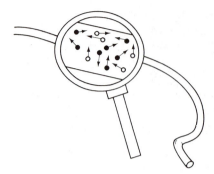

Figure 5-3 Random electron movement. (From T. Weathers and C. Hunter, *Diesel Engines for Automobiles and Small Trucks,* Reston Publishing Company, Inc., Reston, Va., 1981.)

However, what happens if the ends of the wire are attached to the positive and negative terminals of a battery? The negative pole contains an excess of electrons and the positive pole has fewer electrons than it needs to be balanced. The excess electrons at the negative pole repel the electrons in the wire, causing them to flow toward the other end. At the same time, the electrons at the other end of the wire are attracted to the empty orbits at the battery's positive pole (Fig. 5-4).

It is like a game of tag played near the speed of light, where nobody (or no electron) ever catches anybody. The chemical activity inside the battery creates a constant imbalance between the two poles. As long as the battery remains active and the circuit is complete, the electrons will chase each other from empty orbit to empty orbit, through the battery and through the wire. If the battery weakens, the current slows down. If the wire is cut, the flow stops altogether. However, the battery still pulls electrons away from the end of the wire connected to the positive pole and builds up electrons at the end connected to the negative pole.

Electrical flow takes place wherever there is an overabundance of electrons in one place, an underabundance in another place, and an uninterrupted path between the two. The potential for electrical flow exists whenever there is an imbalance but no complete path for current to flow. For instance, even though no current flows when the ignition switch in a car is off, the potential for current flow still exists because there is an electrical imbalance on either side of the switch.

Current flow can be temporary or sustained. The sudden discharge of electricity represented by a lightning bolt is obviously temporary. An electrical imbalance is created between a cloud and the ground. When the potential becomes great enough, the excess electrons are able to create a path through the atoms in the air.

Most sustained current flow is the result of some man-made device: a battery, alternator, generator, and so on. They convert either chemical or mechanical energy into electrical potential. That is, they create regions of electrical imbalance, which, when connected to a complete circuit, will cause electrical flow.

Conductors, Nonconductors, and Semiconductors

The number of electrons in the outer orbit of an atom determines how readily substances containing the atom will allow electrical flow.

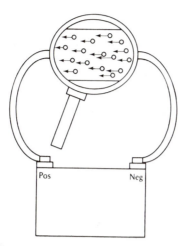

Figure 5-4 Directed electron movement. (From T. Weathers and C. Hunter, *Diesel engines for Automobiles and Small Trucks,* Reston Publishing Company, Inc., Reston, Va., 1981.)

Conductors.   Substances containing atoms with one to three outer orbit electrons are usually considered to be good conductors. The electrons in their outer orbits are loosely held and easily put into motion. Copper, gold, iron, and silver are examples.

Nonconductors.   Materials containing atoms with five or more outer electrons are poor conductors under normal circumstances. Their electrons are tightly held in a pattern with other electrons and are not easily dislodged. Nonconducting materials, such as rubber, glass, and many plastics, are used as insulators to channel and direct electrical flow—making sure that it goes only where it is supposed to go.

Semiconductors.   Semiconductors are very stable elements. Their atoms, containing four outer electrons each, join together in rigid patterns that are not easily disturbed and that do not readily support electron flow. However, this holds true only as long as the element remains in its pure state. If small amounts of impurities are added, the semiconductor becomes a conductor. Depending on the kind of impurity added, current flow will depend primarily on the presence of positively charged "holes" (actually vacant spaces in outer orbits) or on free electrons floating through the semiconductor. Semiconductors are discussed more completely in Chapter 6.

Effects of Electron Flow

Electron flow alone does not have much practical application. The effects of electron movement are what we put to everyday use. For instance, electron flow jostles all the electrons in a circuit. These bouncing electrons cause atoms to vibrate, which results in heat. All current flow involves some degree of heating. If uncontrolled, electrical heating can be dangerous. A wire can melt its insulation and cause a fire. However, by using the right kinds of materials in properly designed circuits, the heat can be put to practical use.

A related effect of the atomic disturbance is light. Every time an electron bounces back and forth between the orbits in an atom, light flashes from the atom (Fig. 5-5). As Thomas Edison proved, by using the right kinds of materials in the right circumstances, practical electric lights are possible.

Besides having an effect on the atoms that make up a conductor, electron flow also causes a change in the surrounding space. When current flows through a wire, revolving rings of magnetic force are created around the wire. These lines of magnetic force, as noted in the second half of this chapter, have many applications, such as in electric control devices.

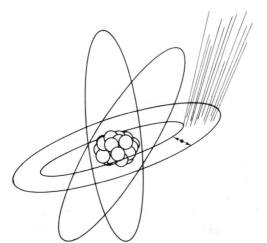

Figure 5-5 Light emission from an atom. (From T. Weathers and C. Hunter, *Diesel Engines for Automobiles and Small Trucks,* Reston Publishing Company, Inc., Reston, Va., 1981.)

Electrical Measurement

We have now seen what electrical flow is, why and how it works, and some of its effects. The last item to be considered before going on to magnetism is electrical measurement.

Volts.   Voltage is a measure of electrical pressure. It is an indication of the pressure exerted on each electron in a circuit by a source of electrical potential. As the following example shows, it is not a measure of the total electrical force felt by all the electrons in a circuit.

To see what this means, imagine you have two 12-volt batteries, a large one that is used in a full-size pickup truck, and a small one that is used in a compact automobile. Both have the potential for exerting 12 volts of force of pressure on each electron in a circuit. In other words, the electrical imbalance existing at the positive and negative poles of both batteries will have the same disruptive effect on a given electron in a given atom. However, the larger battery has the potential for disturbing more electrons and more atoms.

Amperes. Amperage is a measure of total electrical flow. It is an indication of the number of electrons flowing past a given point in a circuit. Amperage (or the number of electrons in motion) depends on electrical pressure (voltage) and on electrical resistance (ohms, discussed next).

Ohms. An ohm is a unit of electrical resistance. It is used to express the varying ability of different materials to support current flow. Some factors affecting resistance are (1) the size of the conductor—just as a larger water pipe will let more water flow than a small pipe, a large conductor will let more electrons flow; (2) the length of the conductor—the longer the conductor, the greater the resistance; and (3) the nature of the conductor—as we have already seen, some materials support current flow better than others.

Ohm's Law

As you might imagine, ohms, amperes, and volts are related quantities. If one quantity changes, at least one of the others must also change. For instance, in a given circuit:

1. If voltage changes (goes up or down) and ohms of resistance stay the same, the amperage must also change (go up or down).
2. If the ohms go up (as a result of a frayed or otherwise damaged wire) and the voltage at the source does not change, the amperage must go down. Not as many electrons can flow.
3. Conversely, if the ohms of resistance drop, the amperage must increase.

These relations are expressed in a statement called Ohm's Law. That law and associated formulas are given in Fig. 5-6.

Ohm's Law:

An electrical pressure of 1 volt is required for 1 ampere of current to flow past 1 OHM of resistance.

Stated as a formula:

$$E = I \times R$$

or

$$I = \frac{R}{E}$$

or

$$R = \frac{E}{I}$$

E = Voltage (or Electromotive Force)

R = Resistance

I = OHMS (or Impedance)

Figure 5-6 Ohm's Law. (From T. Weathers and C. Hunter, *Diesel Engines for Automobiles and Small Trucks,* Reston Publishing Company, Inc., Reston, Va., 1981.)

MAGNETISM

What Is Magnetism?

Magnetism is a mysterious force operating equally well through air, empty space, or solid matter. No one knows exactly how it works. Useful theories have been devised that help predict how magnetism behaves and how it can be put to practical application. We will examine some of these theories because they are needed to help understand automotive electrical operation. Like electricity, magnetism cannot be felt, seen, touched. It is more a description of an effect than of a thing.

What Is a Magnet?

A magnet itself is easier to describe than magnetism. Magnets are objects with the power to attract or repel other magnets and to attract iron and certain materials made from iron. Magnets are directional; that is, they have distinct ends, or poles. If a straight or bar magnet is allowed to hang freely in a horizontal position, one end (called the *north pole*) will always point in a northerly direction and the other end (called the *south pole*) will always point toward the south. No matter how a magnet is cut, shaped, or altered, it (or its pieces) will have two magnetically different poles.

Theory of Magnetic Operation

Scientists believe that magnetic properties originate at the atomic level. What follows is a slightly altered version of the explanation most often given.

Every atom in the universe is thought to spin on its axis, like the earth spins on its axis, or a toy spins about its handle. The spinning action (together with electron movement) causes the atoms to be surrounded by lines of magnetic force. No one knows exactly what these lines of force are, or if they even exist in the usual physical sense. However, for our purposes, we can visualize them as being the paths taken by imaginary particles flying around and through atoms (Fig. 5-7). The particles travel out one end of an atom's axis (its north pole), circle around the atom, reenter at its south pole, then go through the atom to start the trip again.

In most materials, the atomic poles point in different directions. The lines of force surrounding adjacent atoms do not line up; the imaginary particles flying around neighboring atoms bump into one another as often as they travel along parallel paths.

However, something happens in magnetic materials (Fig. 5-8). A substantial number of the atomic poles point in the same direction. This means that a substantial number of the imaginary particles travel in the same direction. Instead of expending their imaginary energy in headlong collisions, the nonexistent particles fly along parallel paths. They join forces to fly from atom to atom, building up enough momentum to actually escape, for awhile, the confines of the magnetic material.

The action duplicates on a large scale what takes place on a small scale within atoms. Imaginary particles fly along parallel paths within a magnet, building up enough power and energy to leave the magnet at its north pole. Then the particles

Atomic
Axis

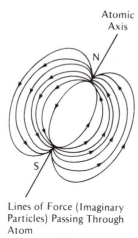

Lines of Force (Imaginary
Particles) Passing Through
Atom

Figure 5-7 Magnetic force lines passing through a simple atom. (From T. Weathers and C. Hunter, *Diesel Engines for Automobiles and Small Trucks,* Reston Publishing Company, Inc., Reston, Va., 1981.)

Random Atomic Alignment

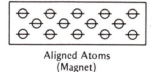

Aligned Atoms
(Magnet)

Figure 5-8 Aligned versus unaligned atoms. (From T. Weathers and C. Hunter, *Diesel Engines for Automobiles and Small Trucks,* Reston Publishing Company, Inc., Reston, Va., 1981.)

circle back to reenter the magnet at its south pole and start the trip again. The paths taken by the particles represent lines of magnetic force, which extend in three dimensions on all sides of the magnet (Fig. 5-9).

The flight of the imaginary particles helps to explain how magnetic materials attract and repel each other (Fig. 5-10). When the opposite poles of two magnets are placed together, the imaginary particles travel in the same direction, just as they do within the atomic structure of the two magnets. Consequently, the particles join forces, pulling the two magnets together. In effect, this creates a single magnet with double the power of the individual magnets.

However, if the like poles of two magnets are put in close proximity, the particles move in opposite directions. As a result, they bump together to push the magnets apart.

Magnets are attracted to iron objects when the lines of force from the magnet penetrate the atomic structure of the iron material. The iron atoms tend to line up with the lines of force, causing the imaginary particles to fly along with same paths. As the particles join forces, they tend to pull the two objects together.

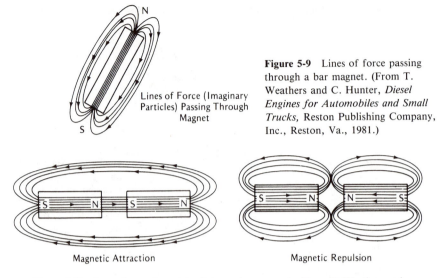

Figure 5-9 Lines of force passing through a bar magnet. (From T. Weathers and C. Hunter, *Diesel Engines for Automobiles and Small Trucks,* Reston Publishing Company, Inc., Reston, Va., 1981.)

Lines of Force (Imaginary Particles) Passing Through Magnet

Magnetic Attraction   Magnetic Repulsion

Figure 5-10 Attraction and repulsion between two magnets. (From T. Weathers and C. Hunter, *Diesel Engines for Automobiles and Small Trucks,* Reston Publishing Company, Inc., Reston, Va., 1981.)

These magnetic flight paths (whatever they represent) can actually be seen by sprinkling iron filings on a piece of paper that has been placed over a magnet. Tapping the paper lightly loosens the filings so that they can follow their natural attraction to the lines of force. The pattern of filings represents the shape of that part of the three-dimensional magnetic field sliced through by the paper. Figure 5-11 pictures the magnetic fields surrounding horseshoe and bar magnets.

Iron Filings Around Horseshoe Magnet

Iron Filings Around Bar Magnet

Figure 5-11 Iron filings around horseshoe and bar magnets. (From T. Weathers and C. Hunter, *Diesel Engines for Automobiles and Small Trucks,* Reston Publishing Company, Inc., Reston, Va., 1981.)

Kinds of Magnets

There are three basic kinds of magnets: permanent, temporary, and electromagnetic.

Permanent magnets. Permanent magnets are often made from the mineral magnetite, which is a naturally occurring magnet. Its atoms are aligned sufficiently to produce a coherent magnetic field. Commonly called *lodestone*, magnetite was discovered hundreds of years ago. Early mariners used it as a crude compass to help chart their course when traveling out of sight of land.

Temporary magnets.   Temporary magnets are created by placing certain iron or iron-based materials in the presence of strong magnetic fields. The lines of force from the magnetic field "pull" the iron atoms into alignment. Depending on the strength of the magnetic field and on the nature of the material, the atoms may remain in alignment after the magnetic field is removed. However, as the name implies, these kinds of magnets may not last very long. Any sort of shock can disturb the fragile atomic alignment; this causes the atoms to become jumbled again and no longer capable of producing a coherent magnetic field.

Electromagnets.   Electromagnets depend on a peculiar property of electricity: whenever current flows through a conductor, the atoms in the conductor line up sufficiently to produce a coherent magnetic field.

If the conductor is a wire, the lines of force form concentric rings around the wire (Fig. 5-12). In other words, the imaginary particles fly in circular paths around the wire. The direction in which the nonexistent particles fly, clockwise or counterclockwise, depends on the direction in which the current is flowing.

Electromagnets are created when current-carrying wires are formed into a coil (Fig. 5-13). This is how it works: Current flows in the same direction in adjacent loops of the coil. Consequently, the imaginary particles fly in the same direction in adjoining loops. As we have seen before, whenever these particles fly in the same direction, they join forces. So the lines of force merge to circle all the loops, traveling around the outside of the coil, going back through the inside of the coil, then going around the outside again.

An electromagnet is like a permanent magnet in many ways. It has a north pole (where the particles come out of the coil) and a south pole (where the particles re-

Lines of Force (Imaginary
Particles) Travel in Circular
Paths Around Wire
Carrying Current

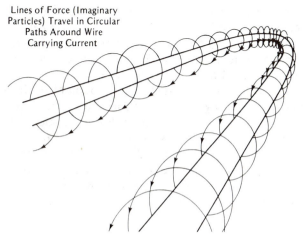

Figure 5-12   Force lines going around wire. (From T. Weathers and C. Hunter, *Diesel Engines for Automobiles and Small Trucks,* Reston Publishing Company, Inc., Reston, Va., 1981.)

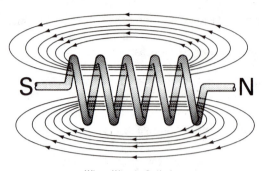

When Wire is Coiled,
Force Lines Merge to
Travel Around and Through
Entire Coil

Figure 5-13 Wire formed in an electromagnetic coil. (From T. Weathers and C. Hunter, *Diesel Engines for Automobiles and Small Trucks,* Reston Publishing Company, Inc., Reston, Va., 1981.)

turn). An electromagnet can attract or repel other magnets. It can also attract iron.

However, there are still some important differences between the two. First, electromagnets can be turned on or off by turning the current on or off. Second, the polarity of the electromagnet (which end is north and which is south) can be reversed by reversing the current flow. Finally, the strength of the magnet can be varied by varying the current flowing through the coil or by increasing the number of loops in the coil.

INDUCED CURRENT FLOW

The relationship between magnetism and electricity is strange and complex. As noted earlier, electrons moving along a wire will create a magnetic field. However, it is also possible to do just the opposite, to create electrical movement in a wire by using a magnetic field.

To see how this works, imagine a horseshoe magnetic being passed over a wire. The magnet's lines of force, as they travel between the poles, cut across and wrap around the wire. The lines penetrate the very atomic structure of the wire. As a result, the electrical balance of the atoms is upset and electrons are put into motion along the wire (as long as the wire is attached to a complete circuit).

The direction of the electron flow is determined by the motion of the lines of force across the wire. If (relatively speaking) the lines of force cut down across the conductor, the current will flow in one particular direction. But if the lines of force move up across the conductor, the current will flow in the opposite direction.

Note: It should be remembered that it does not make any difference if the conductor moves across the lines of force or if the lines of force move across the conductor. All that matters is that there is relative motion between the two. When there is, current flow is induced in the conductor.

CATEGORIES OF CONTROL DEVICES

The principles of electricity and magnetism described in this chapter underlie the operation of an almost unlimited number of devices. For our purposes, they can be grouped into three main categories:

1. *Sensory units.* Converting physical conditions (such as temperature, air pressure, and movement) into electrical current. In computer control systems, these devices are responsible for providing input information. The electrical fluctuations they produce are turned into the raw facts that computers use to make decisions.
2. *Manipulative units.* Converting current flow into physical action. These devices are like the computer's hands, translating its output into changes in ignition timing, fuel delivery, air pump operation, and so on.
3. *Thinking units.* Representing programs or control schemes as electrical patterns. Such devices are the brains of a computer control system.

Sensory units include certain types of inductive generating devices, thermocouples, transducers, and so on. They are discussed in Chapter 10. Manipulative units include solenoids, relays and electric motors. They are discussed in Chapter 11. Thinking units, in other words, the actual computers, are constructed from transistors, diodes, and other components. Transistors and diodes are discussed in Chapter 6, computers in Chapter 8.

6

Semiconductors

In 1956, the Nobel Prize for Physics was awarded to three American researchers Bardeen, Brittain, and Shockley. The first two men invented the transistor and the third, Shockley, explained the physical principles involved. Before that, few people outside the scientific community appreciated the importance of what had been done. Since then, transistors and other semiconductors have been used in everything from portable radios to Apollo moon rockets. Semiconductors have been especially important in the development of computers, ushering in a new generation of lower-cost, powerful, compact equipment. Even before the advent of computerized automotive controls, semiconductors were employed in cars, both in ignition systems and alternator-based charging systems.

To understand automotive systems based on semiconductors, it is necessary to appreciate some of the theory involved. This chapter introduces you to the subject.

ATOMIC STRUCTURE

Semiconductors, as a category of matter, fall somewhere between conductors and nonconductors. The differences are due primarily to the number and arrangement of electrons and to the way in which atoms are joined. The structure of a conductor promotes electrical flow, whereas the bonding between atoms of nonconductors inhibits current movement. Semiconductors may be either conductors or nonconductors, depending on certain circumstances and on the presence of carefully introduced impurities. This "either–or" property of semiconductors makes them valuable as one-way current valves and electrical switches, which, in turn, are used as logic circuits in computers. To see how semiconductors work, it will be necessary to look again at atomic structure, paying particular attention this time to the behavior of outer-shell electrons (Fig. 6-1).

Reviewing from the beginning: all matter is composed of atoms. Atoms, in turn, are composed of electrons, protons, and neutrons. Positively charged protons and no-charge neutrons occupy the nucleus of atoms. Negatively charged electrons move in circular orbits around the nucleus, in much the same way that planets orbit the sun in our solar system. Current flow is a movement of these electrons from one atom to another under the influence of an electromotive force (EMF).

Electrons occupy different orbits or shells around the nucleus. Following a defi-

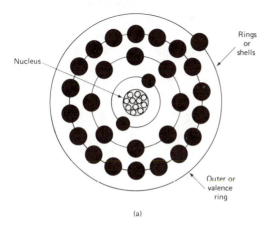

(a)

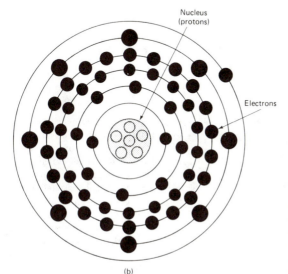

(b)

Figure 6-1   (a) Atomic nomenclature; (b) uranium atom. (From T. Weathers and C. Hunter, *Fundamentals of Electricity and Automotive Electrical Systems,* Prentice-Hall, Inc., Englewood Cliffs, N.J., 1981.)

nite pattern, some electrons orbit close to the nucleus and some orbit farther away. The number of electrons in any given orbit depends on the position of the orbit (first, second, third, etc.).

The electrons in the inner orbits are tightly bound to the nucleus. As a rule they do not enter into reactions with the electrons from other atoms. However, this is not the case with electrons in the outermost or valence orbits. The number of electrons in the valence orbit determines the electrical nature of a substance, whether it will be a conductor of electricity, a nonconductor, or a semiconductor.

CONDUCTORS, NONCONDUCTORS, AND SEMICONDUCTORS

An "ideal" valence orbit contains eight electrons. No valence orbit will contain more than eight, but many will have less. When that happens, an atom will try to "lend" or "borrow" electrons to achieve the satisfied state (Fig. 6-2).

Atoms with one to three valence electrons tend to "lend" electrons. They are conductors of electricity. Their valence electrons are rather loosely held and with a sufficient application of outside energy can be put into motion. For instance, the

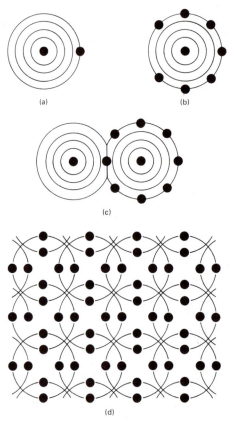

(a)

(b)

(c)

(d)

Figure 6-2 Atoms with valence electrons: (a) atom with one valence electron; (b) atom with seven valence electrons; (c) "shared" valence electrons to achieve "balanced" eight-electron valence orbit; (d) crystal lattice work of semiconductor atoms. Neighboring atoms share four valence electrons to achieve balanced valence orbits. (From T. Weathers and C. Hunter, *Fundamentals of Electricity and Automotive Electrical Systems,* Prentice-Hall, Inc., Englewood Cliffs, N.J., 1981.)

single valence electron of a copper atom can easily be made to drift to the valence orbit of the next copper atom. Then its valence electron will be repelled to the next atom, and so on. The result is electron movement or current flow.

Atoms with five to seven valence electrons tend to "borrow" electrons from other atoms. They are nonconductors. Their valence electrons are more tightly held to the nucleus and cannot easily be put into motion.

The valence orbits of two or more atoms combine to achieve the satisfied valence state. Atoms with one to three valence electrons join atoms with five to seven valence electrons (by the "borrowing" and "lending" process), so the total valence electrons of the combined atoms equals eight. Atoms with four valence electrons combine with one another to achieve the ideal condition. For instance, the four valence electrons of a germanium atom can combine with the four valence electrons of a neighboring germanium atom. The result is a complex latticework crystal of satisfied valence orbits, each having eight electrons.

Because of this structure, atoms with four valence electrons are called semiconductors. In their pure state, considerable voltage or high temperatures are required to break loose electrons from the satisfied valence combinations. However, something dramatic happens to the conductivity of semiconductors when impurities are added. Depending on the kind of impurity, a semiconductor can be made into a carrier of negative charges or positively charged *holes*. The process is called *doping* and the result is a conducting semiconductor. The remainder of this chapter examines the formation and behavior of semiconductors.

Negative, N-type Semiconductors

Negative or N-type semiconductors are created by adding atoms with five valence electrons to the parent material (whose atoms contain four valence electrons). Four of the added electrons combine with four electrons of a parent atom to form a stable, combined valence orbit. However, the fifth added electron has nowhere to go, since no valence orbit can contain more than eight electrons. This electron drifts through the lattice structure of the combined atoms. Under the influence of an EMF, it will support current flow (Fig. 6-3).

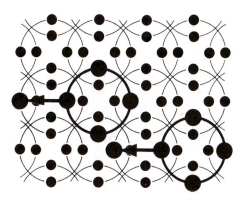

Figure 6-3 N-doped semiconductor. Extra electrons support current flow in the presence of EMF. (From T. Weathers and C. Hunter, *Fundamentals of Electricity and Automotive Electrical Systems,* Prentice-Hall, Inc., Englewood Cliffs, N.J., 1981.)

N-type semiconductors are created by adding materials such as phosphorus, antimony, and arsenic to the parent semiconductor. These additives are sometimes called *N-type doping agents*. Usually, the doping agent is combined with the parent semiconductor at a ratio of 1 atom of doping agent for every 10 million atoms of the parent material.

Positive, P-type Semiconductors

Positive or P-type semiconductors are created by adding atoms with three electrons in their valence orbits (aluminum, indium, boron). These atoms, as before, enter the lattice structure of the parent semiconductor. However, this time, electrons are missing from the valence orbits of the combined atoms. Some of the valence orbits will have only seven electrons instead of eight. The empty spaces in P-type semiconductors are considered to be positively charged "holes" because the "unsatisfied" valence orbits will have a tendency to attract free electrons into the holes, in the same way as if an actual positively charged particle were present (Fig. 6-4).

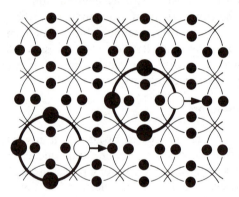

Figure 6-4 P-doped semiconductor. Missing valence electron creates a "positive hole" which will support current flow. (From T. Weathers and C. Hunter, *Fundamentals of Electricity and Automotive Electrical Systems,* Prentice-Hall, Inc., Englewood Cliffs, N.J., 1981.)

CURRENT THEORY: ELECTRON VERSUS HOLE MOVEMENT

To understand semiconductor operation, it is necessary to add to the previously described theory of current flow. Going back to the example of a copper wire attached to the positive and negative terminals of a battery, we have said that the absence of electrons at the positive terminal tends to pull electrons out of the wire and that the excess of electrons at the negative terminal pushes electrons back in to replace those pulled out. The pulling action is due to magnetic attraction between unlike charges, and the pushing action is due to magnetic repulsion between like charges.

If we include the valence theory introduced in this chapter, we can go on to say that the positive terminal attracts the nearest copper atom's single-valence electrons. This attractive force gives the electrons sufficient energy to move from the "balance" energy level up to the conductor. At this energy level, the electrons become free, able to drift under the influence of an EMF toward the battery's positive terminal.

However, something else happens when an electron moves up the conductor level. It leaves behind an empty space or hole in the valence orbit. Since this hole will attempt to capture an electron, it is considered to be positively charged. It will attract negatively charged electrons.

When a shifting electron creates a positive hole, the hole has a tendency to fill itself by attracting another electron from a neighboring atom. Then, when that electron is excited out of its valence energy level, another hole is created. That hole attracts another electron, which creates another hole, which attracts another electron, which creates another hole, and so on. Consequently, as electrons move from the negative to the positive side of the circuit along the conductor level, positive holes will move along the valence level from positive to negative. This means that current flow can actually be described in two ways: as the movement of negative electrons and positive holes (Fig. 6-5).

In the early days of automotive engineering, current flow was usually said to flow from positive to negative (although it was not explained in terms of "holes"). Now, most experts say that current moves from negative to positive. To explain diode and transistor operation, both negative and positive flow concepts are needed.

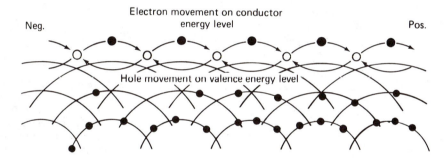

Figure 6-5 Whenever an electron goes up to the conductor level, it leaves a hole behind. This hole attracts a neighboring electron which attracts another electron, and so on. The result is current flow. (From T. Weathers and C. Hunter, *Fundamentals of Electricity and Automotive Electrical Systems,* Prentice-Hall, Inc., Englewood Cliffs, N.J., 1981.)

Current Flow in N- and P-type Semiconductors

Both N- and P-type semiconductors will support current flow. In N-type semiconductors, current flow depends primarily on the movement of free electrons contributed by the doping agent. When a source of EMF is connected to an N-type conductor, the negative side of the circuit pushes electrons through the semiconductor and the positive side attracts the free electrons (Fig. 6-6).

In P-type semiconductors, current flow depends primarily on the empty holes formed by the doping agent. When P-type material is attached to an EMF, the negative side repels the holes.

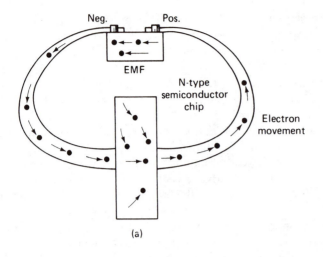

(a)

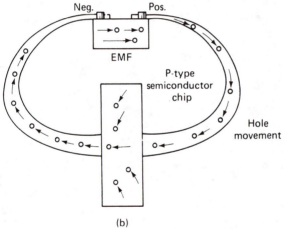

(b)

Figure 6-6 P-doped and N-doped chips: (a) N-type semiconductor in the presence of an EMF; (b) P-type semiconductor in the presence of an EMF. (From T. Weathers and C. Hunter, *Fundamentals of Electricity and Automotive Electrical Systems,* Prentice-Hall, Inc., Englewood Cliffs, N.J., 1981.)

DIODES

In most automotive applications, semiconductors are not used singly. They are combined in two or three layers to form diodes or transistors.

N- and P-type semiconductors joined in thin two-layer chips are called *diodes* (Fig. 6-7). Diodes are used to rectify or change alternating current (ac) to direct current (dc), and special types of diodes help to control the voltage output of alternators.

There are two basic ways in which a diode may be introduced into a circuit. The N half of the diode may be connected to the negative side of the circuit and the P half to the positive side; or the N half may be connected to positive and the P half to negative. Depending on how the diode is connected, it will either allow current to flow or it will act as a barrier to electron movement.

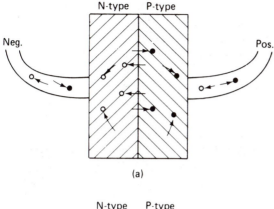

(a)

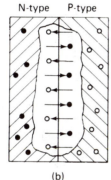

(b)

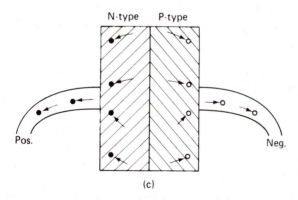

(c)

Figure 6-7 Doides. (a) Forward-biased diode (current flow): electrons and holes cross a P/N junction. (b) Unconnected diode: internal EMF due to cross over at junction. (c) Reverse-biased diode (no current flow): electrons and holes pulled from junction. (From T. Weathers and C. Hunter, *Fundamentals of Electricity and Automotive Electrical Systems,* Prentice-Hall, Inc., Englewood Cliffs, N.J., 1981.)

One key to understanding diode (and later, transistor) operation is to examine the behavior of positive and negative charges at the junction between P- and N-type layers. When a diode is not attached to a circuit, the positive holes from the P side and the negative charges from the N side are drawn toward the junction. Some charges cross over to combine with opposite charges from the other side. However, when the charges cross over, the two diode halves are no longer electrically balanced. In other words, when an electron from the N side goes over to the P side, it leaves a positive charge behind on the N side. The same kind of thing happens when a hole goes from P to N. Consequently, each half of the diode builds up a network of internal charges opposite to the charges at the PN junction. The attraction (or internal EMF) between the opposite charges tends to limit further diffusion of charges across the junction.

When the diode is attached to an external EMF source, the situation changes. If the diode is connected P to positive and N to negative, there will be current flow. The negative pole will push electrons across the barrier as the positive pole pushes holes across. The diode is said to be *forward-biased*.

However, if the diode is attached P side to the negative pole and N side to positive, there will be no current flow. The negative pole will attract the positive holes away from the junction and the positive pole will attract the electrons away. As a result, no charges will cross over the junction. The diode is then said to be *reverse-biased*. These features give the diode the ability to act as an ac-to-dc rectifier or as a one-way current valve.

SEMICONDUCTOR BREAKDOWN/ZENER DIODES

Most charge movement or current flow in a diode is the result of the impurities added to the parent semiconductor. It is called *extrinsic flow* because it is external to or apart from the basic atomic structure of the pure semiconductor. It depends on the extra electrons or holes added by the impurity.

Current flow that depends on the electrons provided by the parent semiconductor itself is called *intrinsic flow* because it is intrinsic or basic to the parent material. Intrinsic flow in most semiconductors is limited to the few electrons than can slip along the atomic cracks and flaws within the crystal structure of the parent material. In other words, these few electrons find pathways through the otherwise satisfied crystal network.

Intrinsic flow in semiconductors increases as the temperature goes up (as opposed to metal conductivity, which increases as the temperature goes down). In most cases, the intrinsic current flow in semiconductors does not become significant until the material nears its melting point—when the atoms are vibrating so much that the valence bonds are about to break apart. However, certain carefully prepared, heavily doped semiconductors can be made to conduct intrinsic current at lower temperatures. (*Note:* The temperature of semiconductors is raised by increasing the voltage impressed against it. The greater the voltage, the greater the force given to electrons moving between atoms and the more the atoms will vibrate. Heat is the result of atomic vibration.)

Zener diodes are constructed from semiconductors which will allow intrinsic current flow above certain voltage/heat levels. Below these levels, a zener diode behaves in a normal manner, allowing only extrinsic, forward-biased current to pass. However, when sufficient, reverse-biased voltage is applied, the diode will "break down" and allow intrinsic current to pass in the opposite direction. Zener diodes are particularly useful in voltage-regulating devices and as protective shunt switches for other circuit components, usually transistors.

TRANSISTORS

Transistors are three-layer semiconducting chips. The two principal combinations are NPN and PNP. In effect, a transistor is made up of two diodes, each sharing a center layer. No matter how the transistor is connected into a circuit, one of the diodes will be reversed-biased and the other forward-biased.

The three layers in a transistor have certain designations (Fig. 6-8). The outside layer of the forward-biased diode (the layer whose polarity is the same as the polarity of the circuit side to which it is attached) is called the *emitter*. The outside layer of the reversed bias diode is called the *collector*. The shared center layer is called the *base*. Each layer—emitter, collector, and base—has its own electrical lead for connecting to different parts of a circuit.

Common materials used for the emitter and the controller are germanium, N doped with phosphorus and P doped with boron. The base section, also commonly made from germanium, is usually only lightly doped, just enough to give a certain minimal number of free charges. The base is very thin compared to the other layers. Its lead is usually attached to a surrounding ring somewhat removed from the emitter–base–collector junctions.

The base section provides a key to transistor operation. To show how it works, Figure 6-9(a) diagrams a PNP transistor with the lead from its base layer connected to an open-circuit leg (which removes the base section from a source of charges). The positive holes in the P-type collector are pushed by the positive charges in the attached circuit to the junction between the collector and the base. On the other side of the transistor, magnetic attraction draws the positive holes in the collector away from the collector–base junction. Many (of the already limited number) of free electrons in the base section are drawn to the emitter–base junction. As a result of this counterbalancing arrangement of charges, few positive or negative charges can pass completely across the base layer. The base layer does not have enough electrons to support anything more than a minor current "leakage" through the layers. Thus the transistor assumes the character of the reverse-biased diode pair.

Current will flow only when the base circuit is completed [Fig. 6-9(b)]. Then it provides sufficient electrons to support hole movement from the emitter across the base to the collector. However, the amount of current flowing through the layers is different. Because the base layer is so thin and because the base circuit is attached to a ring relatively removed from the emitter–base junction, the holes speeding out of the emitter can pass more easily into the collector. Only a limited number of holes will go from the emitter through the base.

The base circuit acts as a control for the emitter–collector circuits. When the

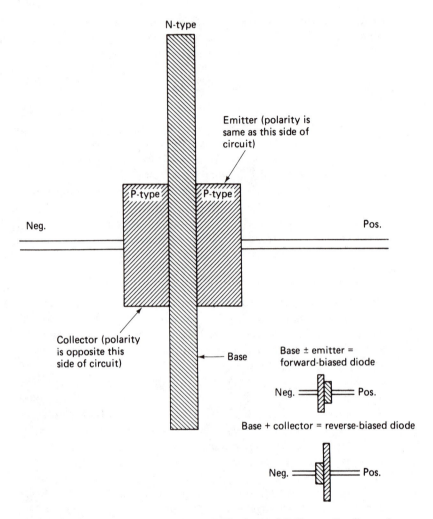

Figure 6-8 Transistor components. (From T. Weathers and C. Hunter, *Fundamentals of Electricity and Automotive Electrical Systems,* Prentice-Hall, Inc., Englewood Cliffs, N.J., 1981.)

base circuit is open, no current passes; and when it is closed, current flows. And because a few charges are allowed to pass through the base, a limited amount of current flow in the base can be used to control much heavier current flows in the emitter circuit. This is particularly useful in the operation of certain transistorized ignition systems, where it is desirable to keep the current flow through the ignition points as low as possible. This feature is also used to fabricate logic *gates*, the basic building block of a computer processor.

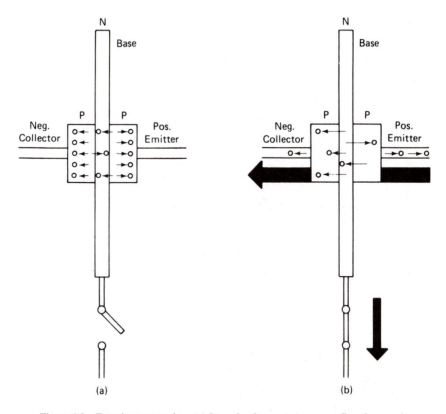

Figure 6-9 Transistor operation. (a) Base circuit open: no current flow from emitter to collector. (b) Base circuit closed: current from emitter to collector. There is a much larger current flow through the emitter–collector than through the base. (From T. Weathers and C. Hunter, *Fundamentals of Electricity and Automotive Electrical Systems,* Prentice-Hall, Inc., Englewood Cliffs, N.J., 1981.)

7

Solid-State Ignition Systems

As of 1977, all American-made automobiles were using solid-state ignition systems. Points-type systems had been eliminated entirely. It is not that points-type systems did not work well; they just did not work well enough in an era of pollution controls and fuel scarcity. Solid-state systems provide higher voltages at the spark plugs and are able to control these voltages better. They also require less maintenance and are less likely to produce fuel-wasting, polluting misfires.

We are reviewing solid-state ignitions for two reasons: (1) they are the direct forerunners of computer-controlled systems, and (2) many of distributor components found in solid-state systems are also used in computer control systems.

CONTACT-CONTROLLED TRANSISTORIZED SYSTEMS

There are a number of devices that come under the general heading of solid-state or transistorized ignition systems. All are similar to points-type systems in these respects: all use ignition coils to create the high-voltage surge needed to operate the spark plugs and all use more-or-less standard secondary circuit components (rotor button, distributor cap, spark plug wires, etc.) to distribute this high voltage. They differ from points-type ignitions basically in the way the primary circuit to the coil is interrupted to produce the high-voltage surge in the secondary.

The first solid-state or transistorized ignition systems were introduced in the late 1950s as optional or add-on equipment for otherwise conventional ignition

74

systems. The primary differences between conventional ignitions of that period and transistorized units are the primary circuit between the points and the ignition coil. In a points-type ignition, current flows more or less directly from the ignition switch, through a registor element (when the switch is in the "run" position), through the closed points, to the primary windings in the ignition coil. When the points open, the flow is interrupted and a high-voltage surge is induced in the secondary windings of the coil.

Primary current flow in a contact-controlled transistorized system does not go directly from the points to the coil (Fig. 7-1). Instead, it goes to the base circuit of a transistor in the amplifier module (Fig. 7-2). The base circuit of a transistor (as you learned in Chapter 6) acts as a switch for emitter and collector circuits. When there is current flow in the base, flow is possible between the emitter and collector. When the base is open, it stops emitter–collector flow. So when the points are closed in a contact-controlled transistorized system, the base allows current to flow from the ignition switch through the emitter and collector to the primary windings of the coil. When the points open, the base circuit is turned off. This blocks current flow through the emitter–collector to the coil. As a result, the coil's primary field collapse and a surge of current is induced in the secondary windings.

Such contact-controlled transistorized ignition systems are considered superior to conventional points-type systems for several reasons: (1) Reduced voltage can be used in the circuit going from the points to the transistor base. This reduces arcing between the points and prolongs point life. (2) Higher-than-normal current can be used in the circuit going through the emitter–collector to the ignition coil. This produces higher secondary voltages and improves spark plug performance in difficult operating conditions.

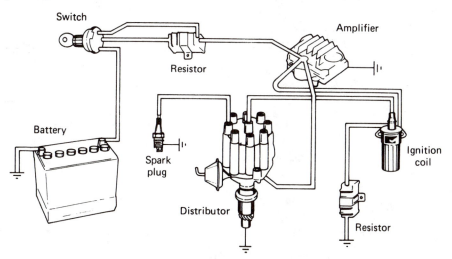

Figure 7-1 Some components in contact-controlled transistorized ignition system. (From T. Weathers and C. Hunter, *Fundamentals of Electricity and Automotive Electrical Systems,* Prentice-Hall, Inc., Englewood Cliffs, N.J., 1981.)

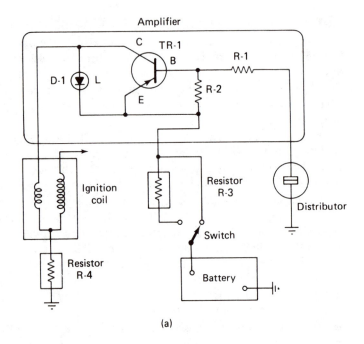

(a)

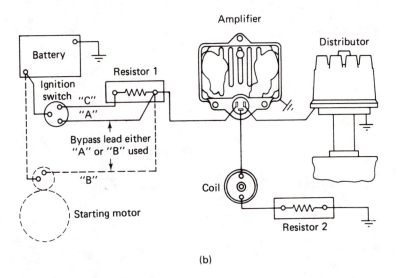

(b)

Figure 7-2 (a) Internal wiring of typical circuit; (b) typical circuit with built-in provision for bypassing resistor during cranking. (From T. Weathers and C. Hunter, *Fundamentals of Electricity and Automotive Electrical Systems,* Prentice-Hall, Inc., Englewood Cliffs, N.J., 1981.)

However, these kinds of transistorized ignition systems still have one fault. They rely on breaker points as the controlling switch. The point assembly, particularly the rubbing block, will wear out in time, even though arcing has been reduced. So manufacturers developed ignition control systems with no breaker points at all. They are called *breakerless* ignitions and eventually came to replace points-type systems altogether.

CAPACITOR DISCHARGE IGNITION SYSTEM

One of the first breakerless ignition systems was the capacitor discharge ignition (CDI) (Fig. 7-3). It replaces the points with a magnetic pulse distributor. This magnetic pulse unit signals a magnetic pulse amplifier (a collection of diodes, transistors, thyrsistors, resistors, and capacitors) to discharge a momentary 300-volt burst of electricity into the primary windings of the coil every time a spark plug is due to fire. When the field created by this pulsing discharge collapses, it induces current flow in the secondary windings to operate the spark plugs.

The main components of the magnetic pulse unit are a timer core, pole piece, pickup coil, and permanent magnet. The timer core, sometimes called a *reluctor*, is attached to the distributor shaft and rotates with it, like the cam in a points-type system. The timer is shaped somewhat like a sprocket, with one tooth for each cylinder. Surrounding the timer core is a pole piece. It also resembles a sprocket, but one that has been turned inside out, with the teeth pointing toward the center. The timer core and the pole piece have the same number of teeth. Whenever the teeth line up, a pulse of current is sent to the pulse amplifier, which in turn signals the capacitor to discharge into the coil's primary winding.

The current pulse is sent to the pulse amplifier because of the arrangement between the timer core, pole piece, permanent magnet, and pickup coil (Fig. 7-4). This is how it works: The permanent magnet is attached to the pole piece. Its magnetic lines of force circle through the pole piece. When the teeth of the pole piece and timer core line up, the lines of force are able to pass between the two and become concentrated. When the teeth do not line up, the lines of force must pass through air and become dissipated. As a result, a pulsating magnetic field exists. It is strongest when the teeth line up and weakest when they do not. These lines of force from the pulsating field cut across the pickup coil. As the lines of force balloon in and out, current flow is generated in the windings of pickup coil. This current flow is not very great, but it is enough to signal the pulse amplifier, which supplies a corresponding but much greater flow of pulsating current to the primary windings of the ignition coil.

Centrifugal and mechanical spark advance are handled in a manner similar to earlier ignition systems. A vacuum advance unit is attached to the pole piece and magnetic pickup assembly. The assembly rotates back and forth in response to changes in manifold vacuum, advancing or retarding the spark as needed. The timer core is attached to centrifugal advance weights and rotates back and forth as engine speed changes. Dwell, however, is fixed, since there is no point gap to consider.

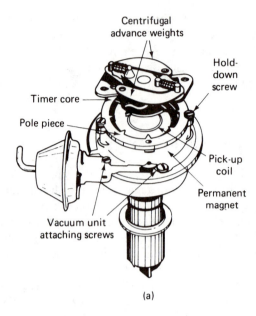

(a)

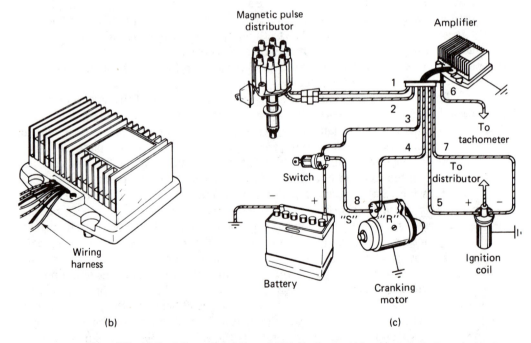

(b)   (c)

Figure 7-3 (a) Partially exploded view of CDI distributor with cap removed; (b) typical ignition pulse amplifier; (c) typical wiring circuit with six-terminal connector on amplifier wiring harness. (From T. Weathers and C. Hunter, *Fundamentals of Electricity and Automotive Electrical Systems,* Prentice-Hall, Inc., Englewood Cliffs, N.J., 1981.)

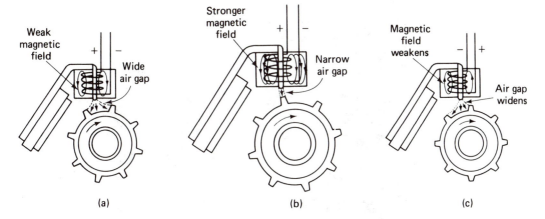

Figure 7-4 Electronic ignition components: (a) Air gap offers resistance to field; (b) increasing field strength induces positive voltage; (c) magnetic field weakens again. (From T. Weathers and C. Hunter, *Fundamentals of Electricity and Automotive Electrical Systems,* Prentice-Hall, Inc., Englewood Cliffs, N.J., 1981.)

OTHER ELECTRONIC IGNITION SYSTEMS

Most current electronic ignition systems are similar (in principle, at least) to the CDI unit just described. All contain a magnetic pulse distributor of some kind and all contain an amplifier unit which controls the primary flow to the ignition coil. The principal difference is the capacitor discharge feature. Most pulse amplifiers do not rely on a pulsating capacitor to operate the ignition coil. Instead, the amplifier unit is somewhat like a (very complicated) switch. It turns off current flow to the primary windings of the ignition coil whenever a signal is received from the magnetic pulse distributor. This momentary interruption in the current flow to the coil's primary windings creates the high-energy surge from the secondary needed to fire the spark plugs.

General Motors High-Energy Ignition (HEI)

This unit (Fig. 7-5) follows the general operating procedures just described. Positive voltage induced in the pickup windings of the magnetic pulse distributor causes the pulse amplifier to be turned on. Then as the polarity of the voltage changes (as the magnetic lines of force balloon in and out), the amplifier is turned off. When the amplifier is turned off, the current flow to the coil's primary side is interrupted. This, in turn, induces a high-voltage surge in the secondary circuit.

Differences between HEI and Chrysler and Ford Systems

The basic principles of a number of Ford and Chrysler systems are similar to those employed in the GM HEI system. However, as noted next, there are several detailed differences.

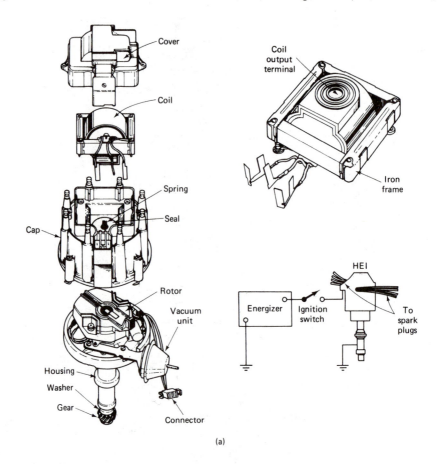

(a)

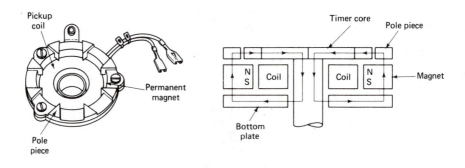

(b)

Figure 7-5 High-energy ignition system: (a) HEI distributor assembly (b) pickup coil. (From T. Weathers and C. Hunter, *Fundamentals of Electricity and Automotive Electrical Systems,* Prentice-Hall, Inc.; Englewood Cliffs, N.J., 1981.)

1. One obvious difference is the placement of the coil. It is located inside the distributor in the HEI system, just beneath the cap. It is located outside the distributor in Chrysler and Ford systems.

2. Another difference is the configuration of the pole piece inside the magnetic pulse distributor. The HEI pole piece surrounds the timer core with an equal number of internal teeth. The pole piece used by Chrysler and Ford has only one tooth or pole piece (Fig. 7-6). However, it works the same as the HEI multi-tooth pole piece. Whenever the single tooth passes a pickup point, a magnetic field is created. As the field balloons in and out, a signal voltage is produced for the pulse amplifier.

3. The HEI pulse amplifier is normally off until a positive voltage pulse from the pickup turns it on. Ford and Chrysler systems, on the other hand, are normally on until turned off by a signal from the pickup unit.

Hall Effect Distributor

The Hall effect distributor (Fig. 7-7) was introduced in 1977 by Chrysler on its Omni line of automobiles. Although using many of the same basic principles as the electronic ignition systems just described, it avoids the problem of signal voltage changing as engine speed changes. As before, the components include a pole piece (reluctor), a magnet, and a pickup unit. However, in this system, the reluctor consists of a number of shutter blades, one for each cylinder. When a shutter blade passes between the magnet and the pickup unit (called the *Hall effect trigger*), the voltage level in a sensor circuit changes. This voltage change is transmitted to the pulse amplifier, which, in turn, operates the ignition coil.

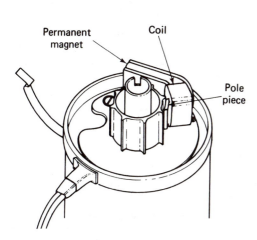

Figure 7-6 Unit with one pickup point. (From T. Weathers and C. Hunter, *Fundamentals of Electricity and Automotive Electrical Systems,* Prentice-Hall, Inc., Englewood Cliffs, N.J., 1981.)

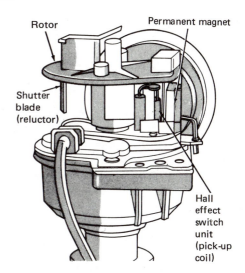

Figure 7-7 Hall effect distributor. (Courtesy of Delco Electronics Division, General Motors Corporation.)

8

How Computers Work

Computers can be approached in two ways: in terms of what they do and how they work. Most users are content to understand just what computers do for them. Usually, there is no need for anyone other than a professional computer engineer to know how the hardware works. As a computer user, you can take this approach and skip this chapter. No automotive technician or owner will ever have to fix a computer. Malfunctioning units are simply replaced. The other chapters explain what computers do in automobiles.

However, if you are interested in some general information about what takes place inside a computer, read on. These basic facts and observations will help you see problems and new developments as parts of a larger picture. You will not have to stay totally in the dark, working strictly on a "monkey-see, monkey-do" basis.

DATA REPRESENTATION

What are computers? As we noted earlier, computers are often described as information processors. However, if you think about it, such a statement is unclear. How can a machine perform a physical action on something as insubstantial as information? Information is what is known, facts and figures. It is not physically real.

The answer is that computers do not process information directly. They process physical conditions that stand for the information. Even though a machine cannot really add 2 and 2 to get 4, it can combine two sets of physical conditions whose total is 4.

Generally, there are two ways to represent information, analog and digital. Automotive systems use both methods. The next few pages examine the differences.

Analog Data Representations

You have already been exposed to analog devices. They are the mechanical and electromechanical control systems described in the first few chapters. Analog devices represent information as measurable shapes, movements, or conditions. For instance, the contour of a camshaft represents valve opening and closing information. Another example is the output of a thermocouple thermometer. It produces voltage fluctuations which represent temperature information (Fig. 8-1). Still another example is the common watch. It portrays time as the movement of hands across a dial face.

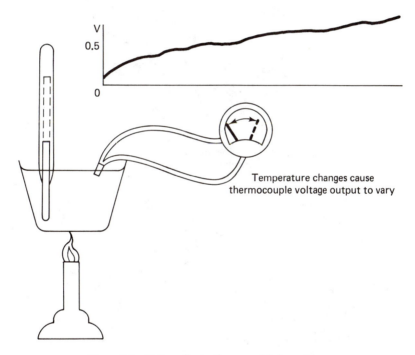

Temperature changes cause
thermocouple voltage output to vary

Figure 8-1 Voltage fluctuations equal information.

In all these devices, the representation is directly analogous to the information—hence the term *analog*. Changes in the position, nature, or quantity of the representation correspond to changes in the information. The representation can be measured directly or observed.

Although Chrysler's first Lean Burn unit was an electronic analog computer, most current systems use analog information primarily for input and output to a digital computer (Fig. 8-2). Here is a summary of what happens:

1. Voltage signals (analog information) from sensors located about the engine are sent to the digital computer.
2. The computer measures the fluctuating voltage signals and converts this information into a set of individual signals or electrical states—in other words, into a pattern of on–off circuits.
3. The information thus represented is altered according to programs.
4. The circuit patterns resulting from the manipulations are converted back into a continuously fluctuating analog voltage output.
5. The output signal is sent back to control devices (solenoids, switches, motors, etc.) which make changes in engine operation.

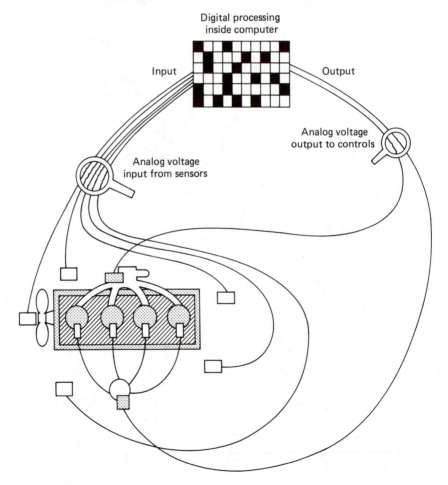

Figure 8-2 Analog and digital information flow.

The final chapters explain the operation of analog input and output devices. The remainder of this chapter will be concerned primarily with the digital operation that takes place inside the computer.

Digital Data Representation

To help see what digital data are, compare a yardstick to the fingers on your hand. The yardstick represents length directly, with each foot and inch merging continuously into the next. It is an analog device.

Your fingers, however, are digital units. In fact, the term *digital* came from the word *digit*, which, among other things, means finger or toe. Each finger can represent a separate, individual quantity or value. Unlike the yardstick, which is a direct analog, or representation of length, your fingers can stand for anything you choose. Normally, they stand for numbers, one finger equaling a count of one, two fingers a count of two, and so on. However, you could just as easily say that two fingers stand for something else.

EXAMPLE OF DIGITAL DATA MANIPULATION

So, on a simplified level, we can say that human fingers comprised the first digital computer. Addition is done by holding up more fingers, subtraction by holding up fewer. Handling numbers greater than 10 is accomplished by assigning a different place value to each finger. In other words, instead of calling the first finger 1, we could say it equals 10, the next finger 11, and so on.

Carrying the example further, we can compare modern digital computers to collections of hands and fingers packed into very small spaces. Such a truly "digital" system might even work, aside from some problems with care, feeding, and worker morale. Before going on, let us see how the design and operation of this imaginary system might be handled. It will be helpful later in understanding the real thing.

Designing a Digital Computer from Digits

First, we need to store information in some physical manner. That is easy in our manual system. We will represent digital information with living digits (Fig. 8-3), just as children do when they compute with their fingers. Patterns of fingers can stand for names, descriptions, and numbers. Addition, subtraction, counting, and other data manipulation will occur as the tiny fingers curl and uncurl.

To keep things from getting confused, the fingers cannot wag about at random; they must operate according to particular directions or programs. Therefore, we will also represent directions for doing things as patterns of fingers. The directions will be broken down into small steps or operations and stored in specified locations called *addresses*. Each element of data that we want to manipulate will also be stored at particular addresses.

Another consideration is efficiency and speed. Certain slow-moving fingers seem

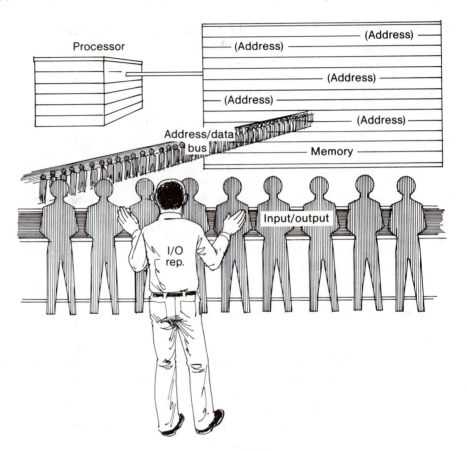

Figure 8-3 Digited digital computer.

better suited to hold information. Other, more nimble fingers are better at manipulation. So we will create two locations in our computer, one called the *memory*, for storing information, and the other, called the *processor*, for doing most of the data manipulation.

Of course, by establishing two regions in the computer, we have also created a communication problem. If you want to add some numbers, you just cannot go to the appropriate memory addresses, cut off some fingers, and move them to the processor. Aside from being messy, it is not necessary. It is the patterns we are interested in, not the actual fingers. So, for internal communication, we will use the hands to which the fingers are attached. We will let the memory fingers transfer their patterns to the palms of the adjacent hands in a kind of braille or touch operation. Those fingers can press on the next palm and so on down the line to the processor region. Also, in the interest of efficiency, we will designate certain hands and fingers for nothing but data transfer. Because they are involved in transportation, we will call them data/address buses.

The final feature we need is external communication. Our computer is no good unless we can get information in and out. For that, we will add some people, calling them input/output (I/O) representatives. They will hold hands with the buses leading to the memory and processor regions. We can transfer data (both facts to be worked on and programs) into the computer by telling an I/O representative to pass it on the appropriate memory address. Output information comes back the same way, but in the reverse direction.

Operating the Digited Digital Computer

Now that we have designed our computer, let us see how it works. Suppose that, for some reason, you cannot remember how to add 2 and 3. However, strangely enough, you do remember how to operate the computer.

First, the numbers must be stored. So you tell the I/O representative to put the number 2 at memory address A and the number 3 at memory address B. The I/O representative says "OK" and in a moment, address A has two fingers sticking up and address B has three fingers raised.

Now you need a program or some directions for performing the operation. Just saying "Add" will not be enough; our computer requires specific directions. It cannot take anything for granted. After thinking for a moment, you come up with these series of steps:

1. Move the pattern (not the fingers) stored in address A to the processor.
2. Move the pattern stored in address B to the processor.
3. Combine the two patterns.
4. Move the resulting sum (or its pattern) to the I/O representative.

This is your computer program. You decided to call it ADD and store the steps at addresses D through H, letting certain patterns of up and down fingers represent each operation. To tie everything together, you also create a master function called CONTROL. Its job will be to make sure that all the program steps are performed at the right time, in the correct sequence. Because it is so important, we will tell I/O representative to store the function directly in the processor.

Now, all you have to do is to tell the I/O representative to execute the ADD program. Here is what happens:

1. The I/O representative informs CONTROL of your intention.
2. Acting like the conductor for the entire system, CONTROL brings the individual ADD steps to the processor region.
3. After arriving along a data bus, the individual steps perform their particular jobs:
   (a) The number patterns are brought to the processor.
   (b) The patterns are combined; in other words, two fingers and three fingers are raised, side by side.
   (c) The new, five-finger pattern is transferred to the I/O representative.

4. The I/O representative says: "Five is the answer." You shake your head and mutter, "Why didn't I think of that?"

MAJOR ELEMENTS IN A COMPUTER

The elements of this imaginary computer are similar, in purpose at least, to the main components in an actual electronic computer (Fig. 8-4). In summary, they are:

1. *Memory:* A collection of electrical patterns that represent stored information.
2. *Processor* (sometimes called the arithmetic and/or logic unit): A group of circuits through which current flows in patterns that correspond to various kinds of data manipulation, such as additions, subtraction, comparison, and so on.
3. *Control:* A particular circuit pattern that manages or oversees the operation of other circuit patterns. Control can be viewed as performing the detailed operations that make up larger operations.
4. *Input/output:* Any device used to move information representations in and out of the computer.

In addition to these similarities, the imaginary computer has other features in common with a bona fide electronic computer. Data and program elements are also stored in memory address. Information representations move along data/address buses. Programs are written to control the way in which patterns are processed.

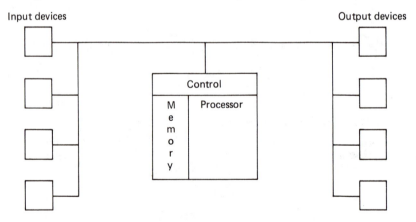

Figure 8-4 Major elements in a computer.

BINARY NUMBERS

One of the primary conceptual differences between our digited digital computer and electronic machines is the number scheme used. The computer made from living digits counts and represents information with the symbols 0 through 9. It requires 10 separate symbols or representations. A mechanical device using this system would also require 10 separate states, conditions, or positions, one for each of the numbers 0 through 9.

Some of the earliest mechanical and electrical computers employed the *base 10* counting system. However, it was awkward to handle so many representations. So most modern digital computers use a *base 2* or *binary* numbering scheme, which requires only two symbols, 1 and 0.

Without getting too deeply into the mathematics of various numbering systems, one way to see how they work is to think about the place or position of a number symbol. Actually, you do it every time you use a decimal number. For instance, consider the decimal symbol 1. By itself, the symbol simply means a count or value of 1. However, put it somewhere else in a group of number symbols and it means something different; for example:

$$1 = \text{one}$$

$$10 = \text{ten}$$

$$100 = \text{one hundred}$$

$$1000 = \text{one thousand}$$

In our base 10 system, everytime you run out of symbols in one column, you carry over to the next column. That is how larger numbers are represented. The base 2 system works the same way. However, since it has only two symbols instead of 10, you have to carry over to the next column more often. Here is a comparison between the representations for zero through 10 in both systems:

Decimal	Binary	Count
0	0	zero
1	1	one
2	10	two
3	11	three
4	100	four
5	101	five
6	110	six
7	111	seven
8	1000	eight
9	1001	nine
10	1010	ten

Binary Difficult for People

As you have probably already decided, binary numbers present certain problems for human use. One problem is the number of places each count requires. It is a lot easier to write 8 than to write 1000. Another unfortunate feature is the choice of binary symbols, 1 and 0. People tend to get them confused with the decimal symbols 1 and

0. As you can see in the table, they are the same only for the first two counts. It would have been better if two entirely different symbols had been selected, + and −, for instance. Of course, the principal difficulty is the unfamiliarity of binary numbers. We grow up thinking in terms of decimal symbols. We visualize them whenever we think of a number. To us, the symbol "3" is the number three. In fact, "11" is just as valid. Both representations are symbols. Neither is the actual number three.

Binary Ideal for Machines

Even though binary numbers present problems for people, they are an ideal way for machines to handle information. For example, suppose that you want to represent numbers using four light bulbs connected by four switches to a power source. With the decimal system, you could represent only the numbers zero through four: zero when all the bulbs are off, one when the first is on, two when the first and second are on, and so on. Using the binary system, on the other hand, you can represent zero through fifteen using four light bulbs. Simply let each off bulb equal 0 and each on bulb equal 1. Figure 8-5 shows how it works.

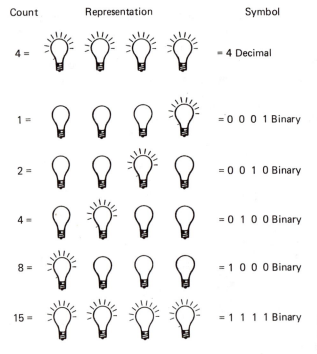

Count   Representation   Symbol

4 =   = 4 Decimal

1 =   = 0 0 0 1 Binary

2 =   = 0 0 1 0 Binary

4 =   = 0 1 0 0 Binary

8 =   = 1 0 0 0 Binary

15 =   = 1 1 1 1 Binary   **Figure 8-5** Light bulbs used to represent data.

Binary Bits, Bytes, and K's

Each of the binary representations is called a *bit*. In a computer, certain collections of bits, usually eight, is called a *byte*. Half a byte, 4 bits, is sometimes called a *nibble*. The capacity of a computer, or the size of its memory, is generally expressed as the number of bits or bytes it can handle. The term K is a shorthand way of stating capacity; 1K equals 1024 units. Therefore, an automotive computer with 48K bytes of random access memory (RAM) can handle more than 48,000 bytes of information.

A PAUSE

At this point we have made all the simple observations. From now on, the path will be tricky. It is like trying to go down two converging roads at the same time, one road called *hardware* and one called *software*. We need to be at the point where the roads cross, where the hardware and the software are the same thing. However, at this level of detail, only limited parts of the operation can be understood. Trying to visualize the entire computer would be like trying to understand the pattern created by a densely packed maze of flickering fireflies—very difficult for most people.

At the other end of the scale, where the roads are far apart, hardware and software are easier to understand. They can be talked about in general terms, like the ones we have been using so far in this book. Looking at a typical operation performed by automotive computers, we might describe a particular piece of software as a program to adjust timing in response to manifold vacuum. At this general level the hardware could simply be passed off as one sensor, one output device, and a computer consisting of a processor, memory, and control. The problem is that the hardware and software are still far apart. We do not know how they work together.

As a practical matter, many computer professionals, especially programmers working on large systems, are not able to put the hardware and software together either. They tend to specialize in one field or another. So, given the fact that even the experts have limited knowledge, it is not realistic to assume that we will be totally satisfied in our attempt to see how hardware and software work together. However, we can get an approximate idea of their relationships.

We will get our understanding by first traveling down the hardware road, then briefly down the software road. After that, at a point where the roads are neither too far apart nor too close together, we will travel along a crossroad to see what connections can be made at that level of understanding. If you have never had any exposure to computers, you will probably need to read this material two or three times before it becomes clear.

HARDWARE ROAD

Bit Representations

The most basic hardware level is the actual representation of bits or 1's and 0's. Virtually anything that can be turned on or off, or that has two states, can represent bits. The four light bulbs noted before are bit representations. So are your fingers. However, modern computers generally represent bits as patterns of high and low voltage, on–off circuits, and regions of opposite magnetic polarity.

Bit Manipulation/Logic Gates

Only three things can happen to bits; they can be (1) left where they are or stored, (2) moved from one place to another, or (3) combined with other bits. Most computer operations, no matter how complex, can be broken down into these basic categories. A moment's thought will also show that it is the last two manipulations—moves and combinations—that form the essence of data processing.

A fundamental tool for manipulating bits is the logic gate. It is an electrical device that produces a given output bit pattern for a given input. Some logic gate combinations are used to store bit patterns; others to perform arithmetic and logic operations, such as addition and comparison.

Computer logic gates can be compared to the switches that control the flow of rail cars through a complex freight yard (Fig. 8-6). Depending on how the switches are set, cars go down this track or that. Sometimes, cars are added to one storage point; sometimes, a car going past one gating switch will cause another car to stop until the path is cleared.

In a railroad yard, the basic units of flow are railroad cars. In a computer, the fundamental flow units are high- and low-voltage pulsations: for example, binary representations of information. High points in the current flow may be described as being positive, 1, or true states. Low points may be referred to as negative, 0, or false conditions. Information is processed or stored when programs specify the paths that high and low pulses take through gates.

Because logic gates are so important, they are sometimes called the *basic building blocks* of a computer. Two of the most important blocks are the AND and OR gates. The next paragraphs describe these gates and their variations.

Kinds of Gates

AND gate. The simplest AND gate has two inputs and one output. The output is high (or a positive voltage pulse) only if both inputs are high; otherwise, the output is low. An AND gate is something like a circuit with two switches wired in series. Both switches must be on for the current to flow. Figure 8-7 shows the symbol for an AND gate and pictures how a comparable series circuit might look. The figure also provides a *truth table* for an AND gate. The table notes all the input and output combinations.

OR gate. This gate works just the opposite of an AND gate. A high signal at either input will produce a high output. As pictured in Fig. 8-8, it is somewhat like a parallel circuit.

NAND and NOR gates. The functions of AND and OR gates can be reversed by a device called an *inverter* or NOT gate (Fig. 8-9). The results are NAND and NOR gates. In a NAND gate, any low input produces a high output. A NOR gate produces a low output for any high input.

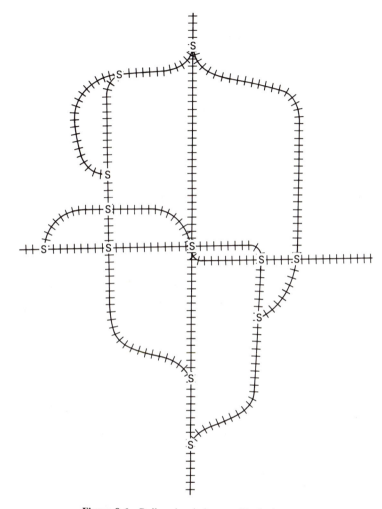

Figure 8-6 Railroad switches act like logic gates.

Gate Circuitry

Gate functions can be obtained from a variety of electromechanical, electrical, and electronic devices. Early computers used relays and vacuum tubes. Most modern computers rely on combinations of transistors, resistors, and diodes. Figure 8-10 shows the basic circuitry for an AND gate. Refer to Chapter 6 for a review of semiconductor operation.

Gate Combinations

Gates are combined in a number of ways so that particular input pulse combinations will produce certain outputs. When the outputs are combined into many levels of

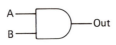

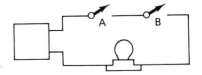

AND gate, like a series circuit

Truth table

A	B	Out
0	0	0
0	1	0
1	0	0
1	1	1

Figure 8-7 AND gate.

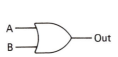

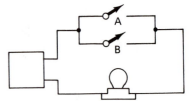

OR gate, like a parallel circuit

Truth table

A	B	Out
0	0	0
0	1	1
1	0	1
1	1	1

Figure 8-8 OR gate.

combinations, data processing results. We look next at two basic categories of combinations: process oriented and storage oriented.

Process-oriented combinations.   These gating devices combine inputs to perform logic and arithmetic tasks, such as comparision, selection, and addition.

Exclusive-Or.   Shown in Fig. 8-11, this combination of AND and OR gates produces a true output only when both inputs are different. The exclusive-or gate can be used in combination with other gates or by itself. One of the operations performed singly is logic comparison. For example, suppose that a program requires

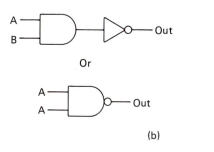

Truth table

A	Out
0	1
1	0

(a)

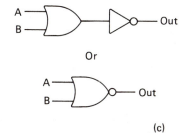

Or

Truth table

A	B	Out
0	0	1
0	1	1
1	0	1
1	1	0

(b)

Or

Truth table

A	B	Out
0	0	1
0	1	0
1	0	0
1	1	0

(c)

Figure 8-9 (a) NOT gate; (b) NAND gate; (c) NOR gate.

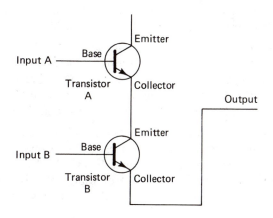

Current must be present at the base of both transistors A and B before any current flow is present at the output of the AND gate

Figure 8-10 AND gate circuit.

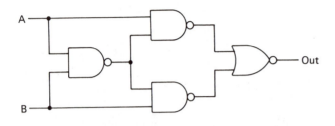

Symbol for exclusive — OR gate

Truth table

A	B	Out
0	0	0
0	1	1
1	0	1
1	1	0

Figure 8-11 Exclusive-or combination.

that a certain action take place only if one other event has occurred and another event has not occurred. Information about the events could be supplied to an exclusive-or gate, which would then make the decision about the desired action.

Full Adder Circuit.   This combination of AND and EITHER-OR gates, shown in Fig. 8-12, can be used to add binary numbers. Individual bits enter at A and B. Carry bits from an earlier addition enter at the "carry-in" point. The carry-in bit is combined with A and B to produce a sum and a carry-out bit.

Decoder.   Created from combinations of AND gates, decoders are used to provide a certain output for a given combination of inputs (Fig. 8-13). For instance, suppose that you want to turn on a switch when a given temperature is reached. Representing temperature information as binary patterns, a decoder could be set to produce a true (positive) output on receipt of the correct bit pattern.

Multiplexers.   A decoder has to examine all its inputs before making a decision about the output. A multiplexer is able, selectively or programatically, to examine one of many inputs. In the illustration pictured in Fig. 8-14, the bit pattern at DCBA determines which input line is examined. A multiplexer can selectively examine the input from a number of sources. Of course, a program has to tell it which and in what order.

Storage-oriented combinations.   As far as the preceding gates are concerned, each set of inputs acts as an individual unit. In other words, the gate combination does not "remember" its own condition from one set of inputs to the next. The de-

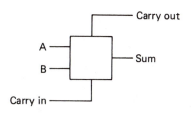

Symbol for full adder

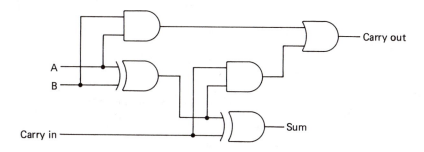

Truth table

A	B	Carry in	Carry out	Sum
0	0	0	0	0
0	1	0	0	1
1	0	0	0	1
1	1	0	1	0
0	0	1	0	1
0	1	1	1	0
1	0	1	1	0
1	1	1	1	1

Figure 8-12 Full adder circuit.

vices that follow remember their own conditions. They act, among other things, as temporary storage units inside the central processor.

Flip-Flops. The basic unit in these combinations is the flip-flop. It is a collection of gates with one or more inputs and two outputs. The outputs are determined by successive inputs. Stated differently, a given input to a flip-flop will affect the output produced by the next input.

Various kinds of flip-flops are used. Figure 8-15 pictures a typical example, called the basic RS flip-flop. As noted on the accompanying truth table, its output changes every time a certain pattern of inputs appears.

Clocked Flip-Flops. Other kinds of flip-flops respond to one or more inputs plus a clock signal (Fig. 8-16). The clock signal is a pattern of current pulses that time or synchronize the actions of many units throughout the computer. Most opera-

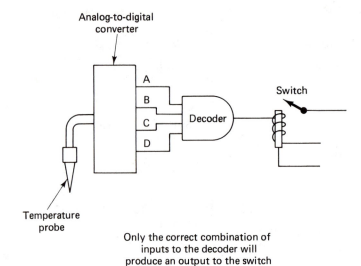

Only the correct combination of
inputs to the decoder will
produce an output to the switch

Figure 8-13   Decoder.

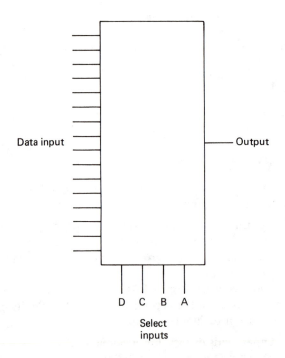

Selection at DCBA determines which data
input will move through multiplexer

Figure 8-14   Multiplexer.

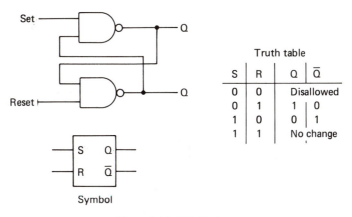

Figure 8-15 RS flip-flop.

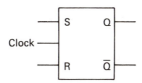

Figure 8-16 Clocked RS flip-flop.

tions require a clock signal so that they will occur in the proper order. Clocked flip-flops respond to other input(s) only when a clock signal is present.

Registers. Just as gates are combined to produce flip-flops, flip-flops are combined to produce registers—another basic computer operating unit. Registers are created from a series of flip-flops acting in a bucket brigade manner. In the example shown on Fig. 8-17, bits are transferred from one flip-flop to the next every time a clock pulse occurs. As noted in the illustration, it takes four counts or beats of the clock to load one 4-bit combination (or nibble) into the register. The nibble is stored in the register until the required combination of inputs occurs again.

Accumulators. Registers used to store the results of a logic or arithmetic operation are called accumulators. Figure 8-18 shows a register/accumulator used in an ADDER circuit. Note that the ADDER has two input registers A and B, an ADD gate combination in the middle, and an accumulator register, C, at the other end. Binary numbers coming from the memory or other processor registers enter the ADD-ER input registers, one at a time. As the numbers are pushed through the input registers, they combined by the ADD device. Upon leaving the device, the results are stored in the accumulator register. These results can now become the input to other circuits.

Basic operating combination. Combinations such as the ones described form the basis of many computer processor operations. Information, in the form of binary bit patterns, is moved to a register, subjected to an arithmetic or logic operation, then moved to an accumulator.

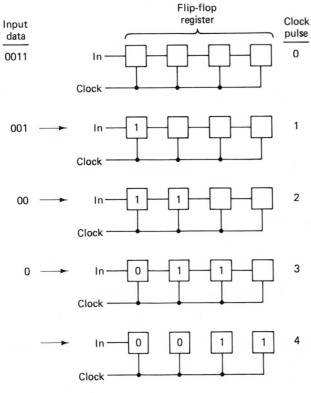

In four pulses of the clock, four bits of
data are transferred into the register

Figure 8-17   Register.

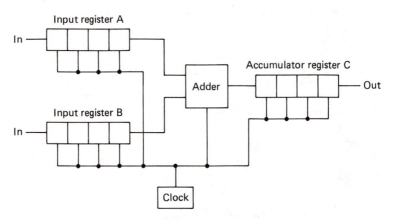

With each pulse of the clock, the contents of
the input registers are moved through the
adder, with the results stored in the
accumulator register

Figure 8-18   Accumulator.

Microprocessor (Computer on a Chip)

Whenever a number of these basic circuit combinations are joined on a single chip or electrical unit, the product is called a *microprocessor*. Basically a computer on a chip, the first microprocessor was created by the Texas Instrument Corporation. Designated the 74181 ALU (arithmetic–logic unit), it could perform all the operations previously handled by computers thousands of times larger. Microprocessors provide the intelligence for many of the "smart" devices produced today: watches, electronic toys, intelligent microwave ovens, calculators—and automotive computers. Figure 8-19 shows a basic microprocessor.

Memory

The one thing most microprocessors lack is extensive memory. Although the registers and accumulators we covered earlier are memory or storage units, their capacity is limited. They are used primarily for temporary storage of working information—facts

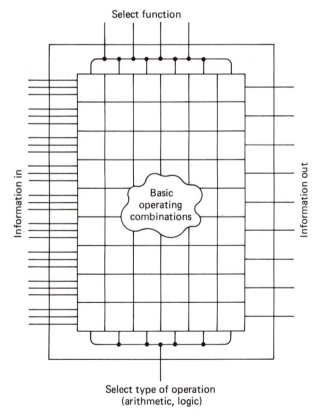

The microprocessor is actually a tiny computer, able to operate a watch, control an electronic toy, time a microwave oven, or help manage a car

Figure 8-19 Microprocessor.

and figures that are being actively processed at the moment. Different types of memory systems are used for storing the bulk of information used by the computer. The type and style of the particular memory system employed depends on the type and nature of the computer.

Large, *mainframe* computers, which must store vast amounts of information, use a hierarchy of memory. Bulk information is usually stored on magnetic tapes and disks, similar to the way audio (analog) information is stored on recording tapes. These devices provide a almost endless repository for information, since they can be added as needed. However, access to the information is relatively slow by computer standards. Therefore, another category of storage, sometimes called *main memory*, is employed. Although the capacity is less than bulk storage, the entry and retrieval rates are much higher.

Main memory in older computers was often formed by lattice-like arrays of tiny doughnut-shaped iron cores. The cores could be magnetized in one direction or another to represent 1's or 0's. Even today, main memory is sometimes referred to as *core memory*. With the widespread application of solid-state circuits, semiconductor gating devices became the primary medium for main memory storage. Today, the most advanced, state-of-the-art machines are now beginning to use patterns of magnetic bubbles in thin wafers of garnet to represent 1's and 0's.

Each of these developments represents a new *generation* of computer technology. However, no matter how they are constructed, or how fast they are, all memory systems serve a common function. They provide an electronic file storage for the central processor, arranging information into the electrical equivalent of cabinets, drawers, and folders.

The memory systems used in the smaller, microprocessor computers found in automobiles are usually of the semiconductor variety. These memory systems are classified into three main types: ROM, PROM, and RAM.

ROM.   The letters ROM stand for *read-only memory*. It consists of fixed patterns of 1's and 0's which can only be read out to the central processor or microprocessor. Once created, no new information can be written into the ROM area.

Some ROMs are created by a technique called *mask programming*. First a semiconductor chip is fabricated with an array of pinpoint transistor or bit locations. Then the chip is coated with a photosensitive film. Shining a light through a masking grid creates a circuit pattern onto the chip. The final step is applying a fine layer of conductive metal to the exposed grid lines. Locations where the grid crosses bit locations become logical 1's; other locations are logical 0's. The pattern thus created represents addressable units of information.

Automotive computers use ROM areas for storing basic operation instructions that apply to the widest possible range of vehicles. The same ROM chips may be used across an entire product line. That is because ROM lends itself to mass production. Although the original setup costs are high, copies of the circuit pattern are inexpensive.

PROM.   These letters stand for *programable read-only memory*. Once created, PROM units act just like ROM. Information can only be read out to the processor; no information can be written in.

The biggest difference between ROM and PROM is the manufacturing method and resultant effect on costs. PROM chips are covered with a maze of fusible links. Using a special PROM burning machine, the links are selectively melted to create the desired bit pattern. This process is especially suited for small production runs. The cost per setup is less than for ROMs, although the cost per unit is higher. Automotive computers use PROMs to store information that is unique for a limited number of vehicle types: particular distributor settings, idle speeds, and the entire range of engine and vehicle calibration settings.

RAM.   So far, we have been talking about memory systems that operate in only one direction, reading out of memory to the processor. Computers that deal with changing, complex information from the outside world also need to be able to add new information, then delete it as conditions change. Automotive computers, for instance, must temporarily store information from all the input sensors, then as the information is used, clear out the storage area for new information. This is the job of RAM, usually called *random access memory* but more properly termed *read/write memory*.

The type of RAM employed in microprocessors is usually fabricated from tiny, semiconductor flip-flop elements located on a chip. As you learned before, each flip-flop output can be 1 or 0, depending on the input. After being set (or reset) the flip-flop bits act just like a ROM or PROM bits. Patterns are read out of the memory addresses to the processor as required. The difference is that the bit patterns can be changed by altering the inputs.

In addition to being used for "writing" as well as "reading" information, another significant difference between RAM and ROM is the duration of stored information. ROM storage is "firmwired" and thus permanent. RAM, on the other hand, is temporary. The tiny switches that make up RAM storage will remain set only as long as power is flowing. Turn the power off and the switch settings, together with the stored information, is lost. Therefore, when you turn off the ignition switch in a car, data supplied by sensors to RAM storage disappears. The only way to get around this situation is to provide an uninterruptible power supply from the car's battery to the computer.

Getting Information into and out of Memory

Individual memory chips can have 16K or greater bit capacity. When combined with other chips, the number gets even larger. Trying to manage the control lines as separate units would be too complicated, even for the complex inner world of computers. Consequently, some of the gating devices that we encountered earlier are used to control the flow of information in and out of memory.

Read controls.   Information read in and out of ROM, PROM, and RAM memory locations is often controlled by a decoder. As you will recall, a decoder produces a given output for a certain pattern of inputs. Suppose, then, that you connected a decoder to a small ROM memory chip (Fig. 8-20). The ROM contains a 4 by 8 grid of bit locations, four on each line. The decoder has four input paths and eight output lines, one for each line of bits in the memory. Now suppose that we call each of the eight lines an address and give it a 4-bit code description. Then, suppose that we design the decoder to produce an ON signal at the appropriate address line whenever the correct 4-bit input pattern occurs. All the processor has to do to read the contents of a particular address is send the correct bit pattern to the decoder.

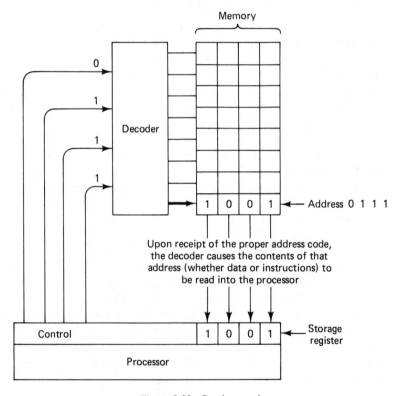

Figure 8-20   Read control.

Write control.   Writing information into a RAM area presents a similar problem. One solution is a series of AND gates connected to the RAM (Fig. 8-21). The output side of each gate is connected to the flip-flop memory elements or latches. The inputs are divided between data entry points and something called *write enable lines*. Whenever an enable pulse and a data pulse occur simultaneously at a gate, an output occurs. The output pulse then sets (or resets) the memory latch to 1 or 0.

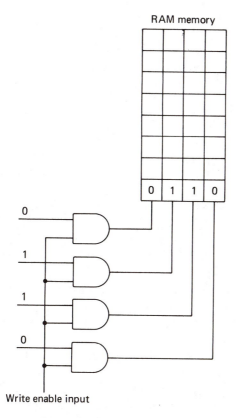

Upon receipt of a write enable pulse,
the AND gates pass data to the RAM   **Figure 8-21** Write control.

Summary of Hardware

We have just about completed our examination of hardware. The study has been rapid and somewhat superficial; however, it is enough to give you an idea of what goes on inside a computer. To help put these components and operations into perspective, let us review their organization, to see what goes into what.

1. The logic gate is one of the most basic elements in a computer. It manipulates bit flow the way switches manipulate rail cars in a freight yard.

2. Gates are joined into two basic kinds of combinations: process oriented and storage oriented.

3. Process-oriented gate combinations include adders, decoders, and multiplexers, to name just a few.

4. The basic storage combination is the flip-flop. Flip-flops are also used in process-oriented devices, although we did not discuss any of these.

5. Various kinds of flip-flops are combined to form registers and accumulators. Registers are working storage units. Accumulators are registers used to store the results of logic or arithmetic operations.

6. Registers, accumulators, and process-oriented gate combinations are joined in still another level of combinations to create basic operating units. Informational bit patterns are brought to an input register (or registers), moved through the process gates, then stored in the accumulator.

7. Basic operating units are joined to create microprocessors—computers on chips.

8. Other than the special-purpose working storage provided by registers and accumulators, most microprocessors do not have much memory. Additional memory chips must be provided.

9. Microprocessor memory systems are generally of the semiconductor variety. Three types are used in automotive computers: ROM, PROM, and RAM.

10. ROM stands for read-only memory. Relatively inexpensive to produce in large quantities, ROMs are used in automotive computers to store general operating information and programs.

11. PROM means programable read-only memory. It acts the same as ROM, allowing information to be read out and not written in. The biggest difference is the cost per setup to produce a particular memory chip. PROMs are less expensive. For this reason, they are used to store basic calibrating information and programs for particular engines and vehicles.

12. RAM refers to random access or read/write memory. RAMs are used to store ongoing information picked up by the sensors located about the engine. They are also used to store the results of calculations performed by the computer.

13. Information can be read out of ROMs, PROMs, and RAMs with the aid of a decoder. Responding to particular address codes, it allows the bits in the corresponding memory address to be moved to the processor.

14. Information is written into RAM addresses in a number of ways. One method employs a series AND gates which, upon receipt of enable pulses, send bits to the desired memory addresses. The pulses set latch, flip-flop elements to the desired pattern of 1's and 0's.

Physical Configuration

The last topic we discuss in the hardware section is the physical appearance of the parts that make up the computer. First, they are small. Individual gates are invisible to the naked eye. A typical microprocessor, which includes hundreds or thousands of gates, might be no larger than 1/2 inch square.

The devices are also integrated. The transistors, diodes, resistors, and other components that make up a chip are permanently bonded together. There are no moving parts, except for the electrons that flow through the circuits.

Chips themselves are assembled on boards. Printed-circuit paths connect the components. Some automotive computers have three separate boards: one for power, one for the processor and memory, and another for the I/O controls.

If testing reveals a computer failure (as discussed in Chapter 13), the entire unit is usually sent back to the factory. The inner workings are generally inaccessible for local repair. However, one of the GM computers does allow replacement of a PROM chip to change engine calibration settings.

Although not part of the computer itself, input and output devices are part of the computer control system. These devices, which are more familiar than computers to most mechanics, are described in the following chapters.

SOFTWARE ROAD

Now it is time to examine briefly the software directions that tell the equipment what data to manipulate and which operation to perform. Two convenient categories for grouping software are as descriptions and as physical states. Within the categories are a number of levels.

As Descriptions

Until it is incorporated into the hardware, the software is actually a description of what takes place. The most general level of software description is a simple statement of the task to be performed. Such a statement, as noted earlier in this chapter, could be "a program to adjust ignition timing in response to changes in manifold air pressure." This is easy enough to understand, but, as we noted before, does not tell us much.

The next level of description is a detailed statement of the task. For instance, a designer might say that if such and such conditions meet certain criteria, a particular action will occur in a given manner. These statements, which often include a number of "if–then" operations, can be understood more easily in flow chart form.

Such software descriptions are all that the mechanic or mechanically inclined owner are ever likely to encounter. The last level of description belongs to the programmer. Programmers write or devise programs, the detailed, technical descriptions of the information management operations. At the creation stage, or anywhere outside the computer, programs are descriptions only.

Programs are stated in various ways. Some are expressed directly in the terms of the machine, as 1's and 0's. Most, however, are written in one of the computer languages: COBOL, ALGOL, BASIC, FORTRAN, PL/1, and so on. These languages, which are actually sophisticated computer programs, convert certain statements into the 1's and 0's required by the machine.

As Physical States

Once converted (or compiled, interpreted, or assembled), programs become a pattern of 1's and 0's inside the computer. In a manner of speaking, the software becomes hardware. In fact, the programs existing in ROM and PROM are sometimes termed *firmware*.

Two main levels of instructions exist in the hardware (Fig. 8-22). One, the *macro level*, corresponds to the most basic steps devised by the programmer, or in some cases by the language program. Macro instructions specify instructions to be performed and the address or location of the individual unit of data that is to be manipulated.

The presence of a macro instruction in the processor automatically initiates a series of micro instructions. Micros are the lowest level of program operation. Remaining all the time in the processor, micros manage the flow between gates, registers, accumulators, and so on. The programmer generally does not have to specify which micros are to be performed; they are an automatic result of the macro operation.

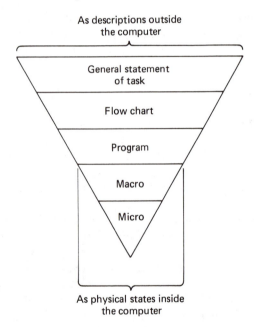

Figure 8-22 Hierarchy of software.

CONNECTING ROAD

At this point, you may be somewhat confused, especially if this is your first exposure to computers. You have encountered a number of new ideas and terms; trying to keep them all organized is not easy. To help in this task, we will, as promised, take a connecting road between the software and hardware paths traveled earlier. We will not try to tie all the information together, just enough to give a coherent picture of the operation of a small computer.

The computer we examine, as pictured in the next illustrations, will be our vehicle for understanding. It does not correspond to any computers that actually exist, but is a composite device incorporating some of the basic features found in various systems.

Quick Trip

First, let us take one fast run down the connecting road to get an overall idea of the country being traveled:

1. Looking at Fig. 8-23, notice that our hypothetical computer has the same major components noted before:
   (a) Input and output devices
   (b) Input and output controls, including analog-to-digital and digital-to-analog converters
   (c) ROM, PROM, and RAM memory elements plus their control devices
   (d) The main processing area with data and instruction registers, microinstruction controls, arithmetic–logic units, a clock, and output accumulator registers
   (e) Two buses or transmission lines, one for data and instructions and the other for micro-control signals

2. When the engine starts, certain basic programs or steps are brought out of the ROM into the processor area. These programs, which control such vital functions as ignition timing, fuel delivery, air pump operation, and so on, are cycled through the arithmetic–logic units one step at a time, macro by macro, micro by micro. However, as far as the operator is concerned, there is no perceptible lag between steps since the computer operates so fast, performing many steps each second.

3. The information that the programs need comes from sensors located about the engine. After moving through the I/O control area, and in some cases being converted from analog-to-digital form, most information is stored momentarily in the RAM area.

4. When called for by particular micro controls, blocks of information are moved out of the RAM addresses to the processor.

5. Information from the sensors, once in the processor, is subjected to various logic and arithmetic operations. Basic calibration facts stored in the PROM regions are also brought into the processor as needed.

6. At certain points in the program steps, the relationships between incoming facts are examined. In other words, the processor, using its process gates, asks if this number is larger than that, less than another, and so on. Depending on the answers to these questions, certain output signals are sent to the I/O control. These signals, which may be converted back to analog form, go to the devices that directly manipulate the operation of the engine: for example, distributor, fuel injector, air pump, and so on.

7. Such basic operations occur many times a second, whenever the engine is run-

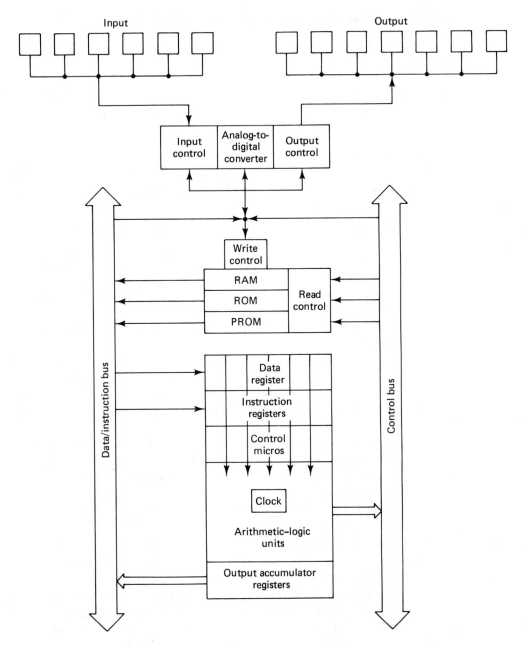

Figure 8-23 Imaginary small computer.

ning. As the input information varies, the output is adjusted.

8. Other actions take place on a nonperiodic or on-demand basis. For instance, some cars provide dashboard displays of engine and trip-related information whenever the operator presses a particular button on the dash.

9. This information, including average miles per gallon, average speed, and distance to a certain point, is provided by special programs. These programs can also be located in ROM or PROM and brought into the processor region as needed. The information required by the programs can also be stored in RAM.

A Slower Trip

Now let us go down the connecting road again. This time we will follow the flow of one segment or "slice" of information from the manifold air pressure sensor (Fig. 8-24). The flow from other sensors proceeds in somewhat the same way.

1. Responding to changes in manifold pressure, the sensor produces a varying voltage signal.

2. Looking at one segment or slice of the voltage signal, it goes first to the I/O control area. There the signal is converted from analog form to digital form. Our slice of information becomes a pattern of 1's and 0's that stand for the voltage level coming out of the sensor.

3. The unit of information is routed through the write control to a particular address in the RAM storage area.

4. Within a few pulses of the computer clock, a macro instruction dealing with manifold air pressure moves into the instruction register. The request comes along the control bus from the processor. The instruction then goes down the address/data bus to the register.

5. The presence of the macro in the register initiates the particular micro instructions that make up the macro.

6. One of the micros tells the read control to send our slice of information down the data bus to one of the processor data registers.

7. Now a second slice of information from another sensor is moved into another processor register. The two slices, proceeding bit by bit with each clock pulse, are passed through the arithmetic–logic unit. In this imaginary example, we can say that they are added together and the results put into the output accumulator register.

8. The combined slice of information then moves to RAM address, to be stored for a moment while other information passes through the system.

9. After a time, a new instruction calls for the combined slice, plus another segment of information from the engine calibration values stored in the PROM region. The two units are passed through a series of process gates to see which is larger.

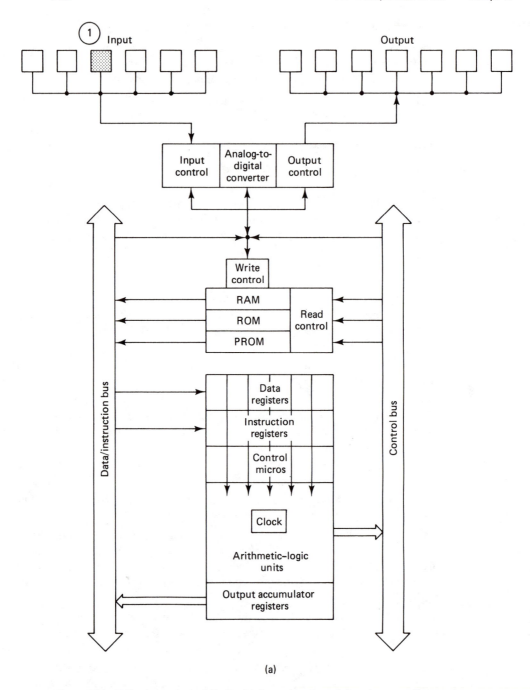

(a)

Figure 8-24 Following a single "slice" of information through imaginary small computer.

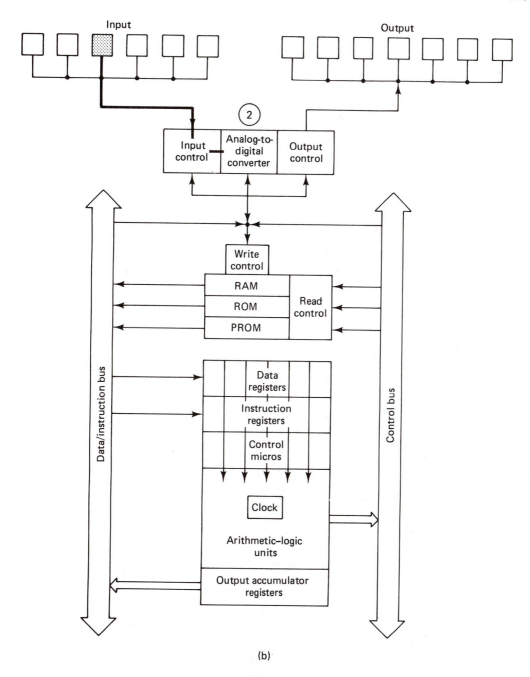

(b)

Figure 8-24 (*continued*)

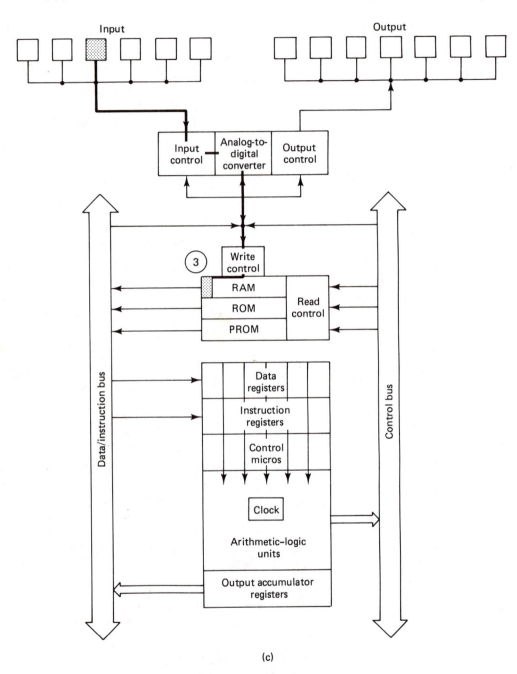

(c)

Figure 8-24   (*continued*)

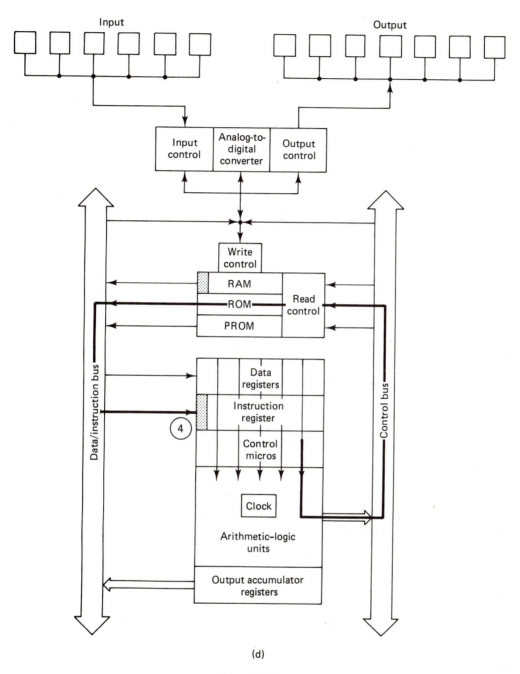

(d)

Figure 8-24 (*continued*)

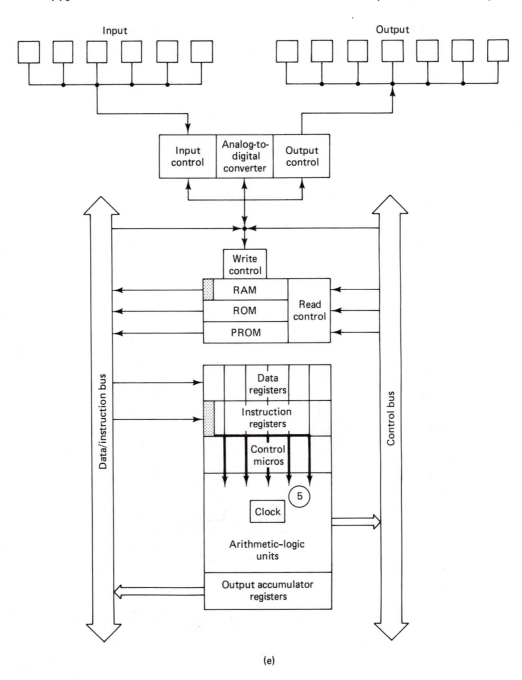

(e)

Figure 8-24   (*continued*)

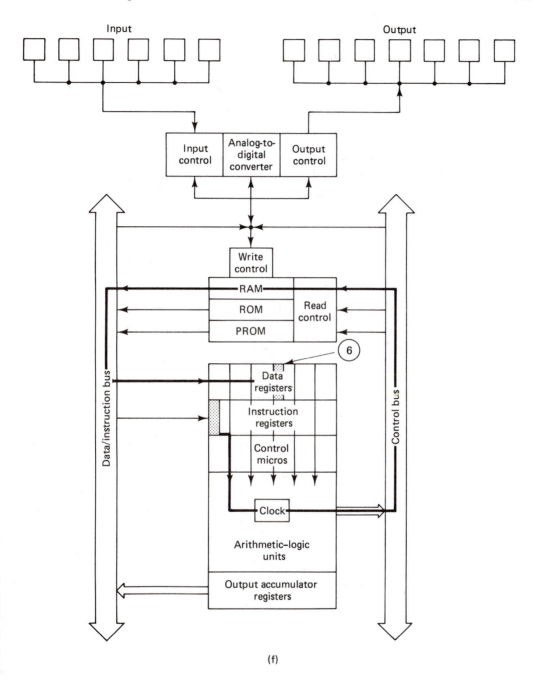

(f)

Figure 8-24 (*continued*)

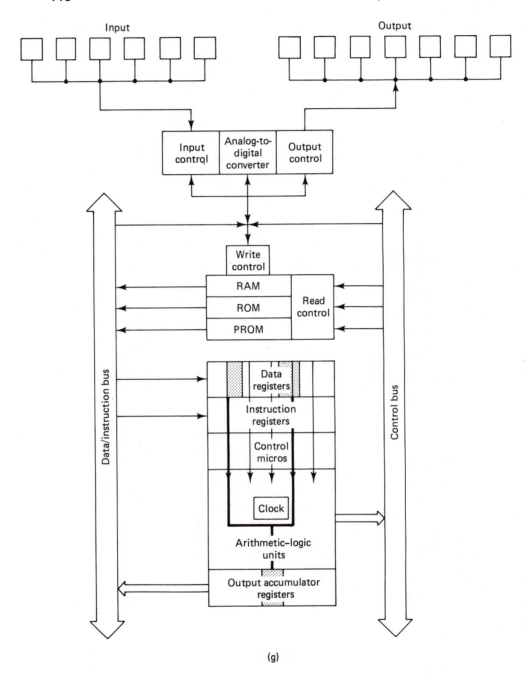

(g)

Figure 8-24   (*continued*)

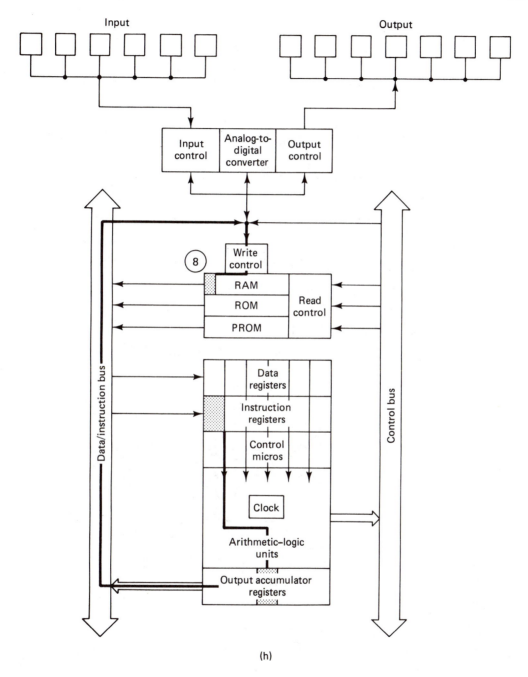

(h)

Figure 8-24 (*continued*)

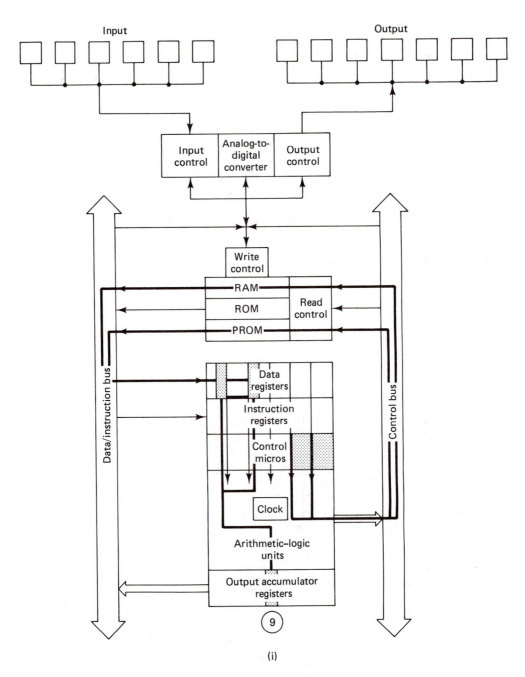

Figure 8-24 (*continued*)

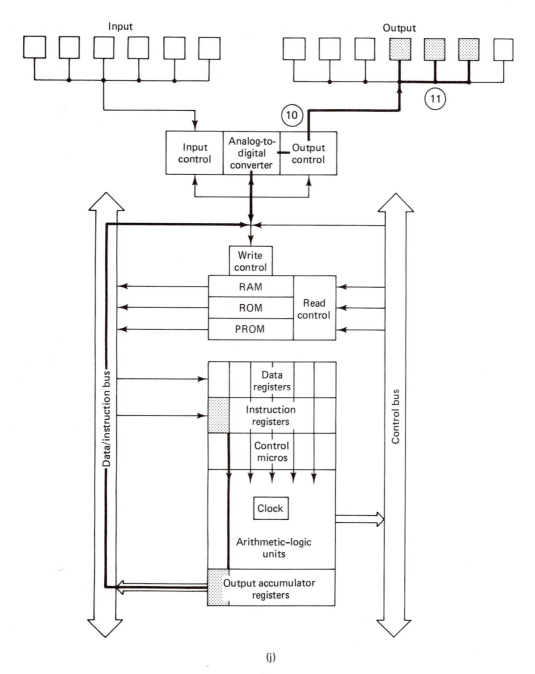

(j)

Figure 8-24 (*continued*)

10. Finally, the much modified slice of information is processed by an operation whose result is an output signal at the I/O control.

11. This signal, the last vestige of our original manifold pressure information, is converted back into analog form. The resulting voltage signal is routed to one of the output controls on the distributor, fuel injector, air pump, or any other device requiring adjustment as a result of the momentary fluctuation in manifold air pressure.

9

Computer Functions

Chapter 8 introduced you to the operation of computers. This chapter looks at some of the major functions performed by automotive computers. Representative systems produced by four manufacturers will be examined:

1. GM Computer Command Control: illustrating a complete digital control system
2. Chrysler combustion control computer: illustrating a system combining digital and analog controls
3. Bosch Motronics: illustrating a European, digital, original equipment manufacturer (OEM) system
4. Ford electronic instruments: illustrating a digital dashboard system

This list is by no means complete. For instance, Ford also produces a complete digital control system similar to the one used by GM, and GM produces electronic instruments. However, our purpose is not to examine completely every system produced by every manufacturer. That would be impossible in a single textbook. Instead, we are interested primarily in obtaining an overview of major computer control functions.

WHAT IS A COMPUTER FUNCTION?

Before getting into the various systems, it will be helpful to arrive at a definition of the term *computer function*. Obviously, it has something to do with what a computer does. Adjusting spark timing in response to manifold pressure is certainly a computer function. However, such statements, as indicated in Chapter 8, are also aspects of computer software. They are general descriptions of computer programs, which, when moved into computer processor, form the basis of all computer operations. So, whenever you read about a function doing this or that in response to one input or another, remember: the description actually refers to the movement of data representations inside the computer. The statement is a layman's way of stating the computer programs.

GM ELECTRONIC CONTROL SYSTEMS

Survey of Functions

Called CCC for Computer Command Control, the GM system provides programs for managing a number of operations and devices (Fig. 9-1). Although GM sometimes describes these operations as if they are each a stand-alone system, it should

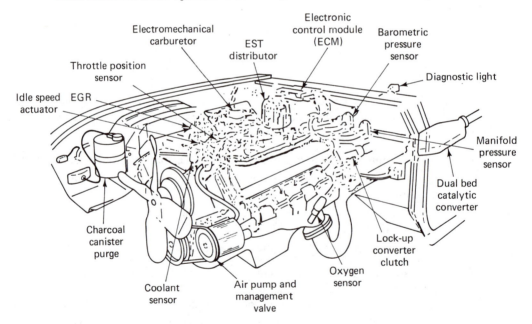

- Model years 1980½–1981
- Closed-loop carburetor control (CLCC)
- Electronic spark timing (EST)
- Idle speed control (ISC)
- Exhaust gas recirculation (EGR)
- Dual-bed catalyst air management system
- Canister purge
- Torque converter control (TCC)

Figure 9-1   GM Computer Command Control system (carburetor version). (Courtesy of General Motors Corporation).

be understood that all are really part of a single computer-controlled system. The particular functions include:

1. Carburetor/fuel injection
2. Spark timing
3. Idle speed
4. Exhaust gas recirculation
5. Canister purge
6. Cruise control
7. Air management
8. Transmission torque converter clutch
9. Early fuel evaporation
10. Modulated displacement (engine size)
11. Instrument panel display
12. System self-diagnosis

Not all of these operations are controlled on all models on all engines. Some, like item 10 (the V-8, V-6, and V-4 cylinder selection feature on large Cadillac engines), may be dropped altogether as new engines and new models are introduced. However, it is likely that most of these functions as well as functions not even introduced yet will become standard features on all models before too many years have passed.

ECM Computer

The brain of the system is the ECM or Electronic Control Module. It is a small digital computer, similiar, in theory at least, to the one we examined in Chapter 8. The ECM contains the control logic for the entire system (Fig. 9-2). Basic operating programs are stored in the ROMs. Vehicle and engine calibration information is maintained in plug-in PROM chips. RAM storage, as noted in Chapter 8, is used as a temporary repository for ongoing operating information and calculations.

The next sections examine the 12 functions in the GM system.

Function 1: Closed-Loop Digital Fuel Injection (CLDFI)

This function (actually a set of programs in the ECM) controls the operation of two fuel injectors located in a throttle body mounted on the intake manifold. The programs cause a signal to be sent to each injector telling it when to open and how long to remain open.

Closed loop.   The term *closed loop* refers to the feedback relationship between the input devices, particularly an oxygen sensor in the exhaust, and the fuel injectors (Fig. 9-3). During normal running, injector operation is based in large part on feedback from the oxygen (O_2) sensor. The O_2 readings, on the other hand, relate to the amount of fuel delivered by the injectors. So the injectors and input devices

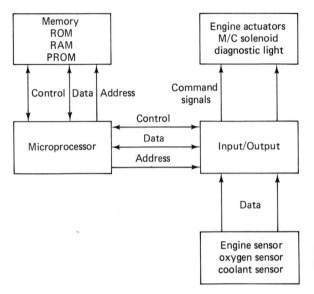

Figure 9-2 ECM computer schematic. (Courtesy of General Motors Corporation.)

Dual-bed catalyst air management system

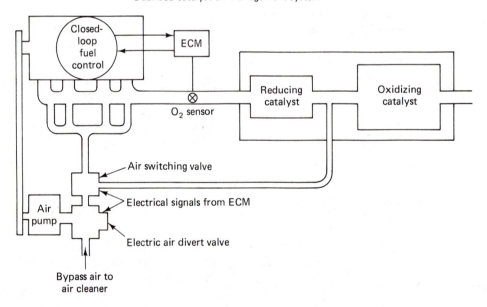

Figure 9-3 Closed-loop operation. (Courtesy of General Motors Corporation.)

operate in a closed-loop fashion, each affecting the other in a relationship determined by the programs running in the ECM.

Throttle body.   A throttle body is what it says it is, a device containing air intake throttle plates. As in a conventional system, the position of the throttle plates determines how much air goes into the engine. However, instead of a carburetor, the throttle body has two solenoid-actuated injectors (Fig. 9-4), one over each plate (action of injectors explained in Chapter 11). Such fuel-injection systems are different from units that position one injector near the intake port of each cylinder. Throttle body injection uses fewer components and is, in principle, less expensive.

Once the atomized fuel reaches the injector plates, the fuel flow through the intake manifold is similar to the flow in carburetor-based systems. In fact, many computer-controlled fuel systems employ carburetors. They are similar to conventional carburetors with the exception of a solenoid-actuated metering rod controlled by the computer. However, it is possible that the superior control provided by throttle body injection will eventually result in the replacement of carburetors: computer controlled and otherwise.

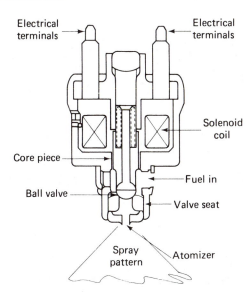

Figure 9-4   Cross section of fuel injector. (From T. Weathers and C. Hunter, *Fundamentals of Electricity and Automotive Electrical Systems,* Prentice-Hall, Inc., Englewood Cliffs, N.J., 1981.)

Reasons for closed-loop system.   The throttle body, closed-loop system was developed in conjunction with the *three-way catalytic reactor*. The first catalytic reactors, introduced in 1975, used two main catalytic components, platinum and palladium. As explained in Chapter 3, these components helped reduce the content of HC and CO contained in the exhaust. The new reactors have added a third ingredient, rhodium. It helps eliminate NOx. Heretofore, the primary means of lowering the exhaust content of the pollutant was the EGR valve.

The problem with the third ingredient is the effect it has on fuel mixture requirements. The first two components, platinum and palladium, require a mixture of 14.7:1 or leaner to be effective. This requirement resulted in several years of lean-burning engines. Chrysler's Lean Burn system, the first electronic engine control, was developed for this reason. Zirconium, on the other hand, must operate on mixtures of 14.7 or richer. Consequently, for all three catalysts to work satisfactorily, the mixture must be very near the stoichiometric ratio, 14.7:1 (Fig. 9-5). The only way to achieve this ideal, very exacting level of control is with an on-board computer.

Input.   The CLDFI programs obtain information from these units (Fig. 9-6):

1. Coolant sensor
2. Manifold air temperature sensor (MAT)
3. Manifold absolute pressure sensor (MAP)
4. Barometric pressure sensor (BARO)
5. Ignition switch (crank and run positions)
6. High-energy ignition module (HEI)
7. Throttle switch (open or closed)
8. Throttle position sensor (angle of throttle plate)
9. Oxygen sensor
10. Fuel pressure sensor

Output.   Output control information is sent to the two throttle body fuel injectors.

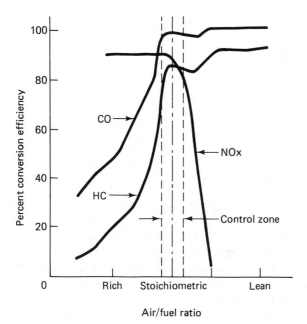

Figure 9-5   Operating band of three-way catalytic reactor. (Courtesy of General Motors Corporation.)

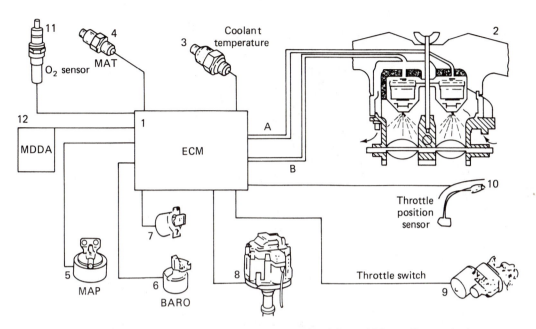

Figure 9-6   CLDFI components. (Courtesy of General Motors Corporation.)

Operating modes.   The closed-loop fuel injection subsystem has eight major and minor operating modes. The particular mode is decided by the programs running in the ECM, based on information supplied by the sensors. These modes or methods of operation address the following major engine conditions:

1. Starting
2. Normal running
3. Cold engine running
4. Altitude variations
5. Idle mixture variations
6. Acceleration
7. Wide-open throttle
8. Deceleration

Starting.   As soon as the ignition switch is turned from the OFF to the ON position, the following sequence of events takes place:

1. Power is supplied to the ECM.
2. The computer's internal clock is reset to zero.
3. The starting programs start to cycle through the computer processor.

As noted in the program flowchart pictured in Fig. 9-7, one of the first program steps is to turn the fuel pump on so that the fuel system will become pressurized. After that, the programs check for a signal indicating that the key switch has been turned to the CRANK position. If a signal is not received within 1 second, the pump is turned off. The pump remains off until the crank signal is received.

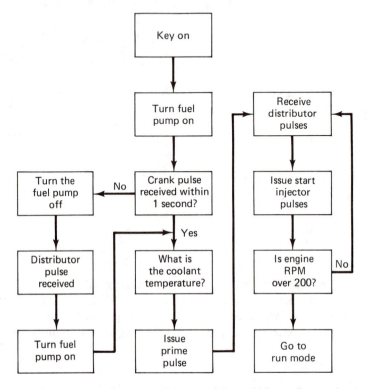

Figure 9-7 Starting flowchart. (Courtesy of General Motors Corporation.)

After the crank signal arrives at the computer, the programs send a primary pulse to the injectors. Upon receipt of the impulse, both injectors spray fuel into the intake manifold. The duration of the signal, and hence the amount of fuel sprayed, depends on the reading from the coolant sensor. Cold engines receive sprays lasting up to 170 milliseconds; warm engines get sprays as short as 10 milliseconds.

After the initial priming pulse, injector operation is timed by reference pulses from the distributor (Fig. 9-8). These reference pulses are associated with the spark timing period.

During the cranking period, both injectors spray fuel with each distributor reference pulse. The duration of the injector signal is again determined by the temperature of the coolant (Fig. 9-9). The lower the temperature, the longer the injectors remain open. (Of course, the injectors cannot remain open longer than the interval between distributor reference pulses. If that happened, the injectors would stay open

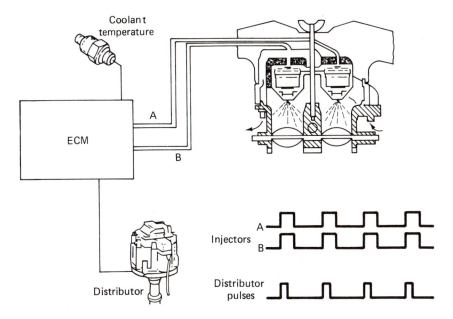

Figure 9-8   Cranking fuel control. (Courtesy of General Motors Corporation.)

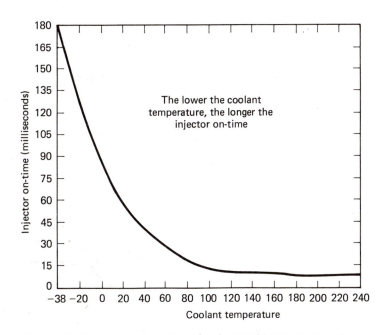

Figure 9-9   Injector on-time. (Courtesy of General Motors Corporation.)

all the time.) During the entire cranking period, engine speed is monitored. When it exceeds a preset PROM value, usually 200 rpm, the next operating mode enters the picture.

Normal Running. During normal running, the system operates in true closed-loop fashion (Fig. 9-10). The injector solenoids are now energized alternately. With each distributor reference pulse, one injector opens; on the next pulse, the other injector opens. The mixture is controlled at the ideal 14.7:1 ratio by varying the pulse duration. Input factors affecting the injector on-period during normal running include (1) manifold air temperature, (2) manifold air pressure, (3) fuel pressure, and (4) O_2 content in the exhaust. Once inside the computer, this information is compared to preset PROM values. Injector operation depends on the result. The remaining DEFI operations can be looked on as special situations occurring during the normal mode.

Cold Engine Running. As in a precomputerized engine, a choke factor is needed to supply extra fuel when the temperature drops below a certain point (Fig. 9-11). Otherwise, not enough fuel will be vaporized, resulting in a mixture leaner than the desired 14.7:1 stoichiometric ratio.

The choke program examines inputs from the coolant sensor, the manifold air pressure gauge, and an elapsed time counter built into the computer. Until the engine reaches its normal operating temperature, the choke program causes the injectors to remain on longer. Like precomputerized choke systems, the program modifies the

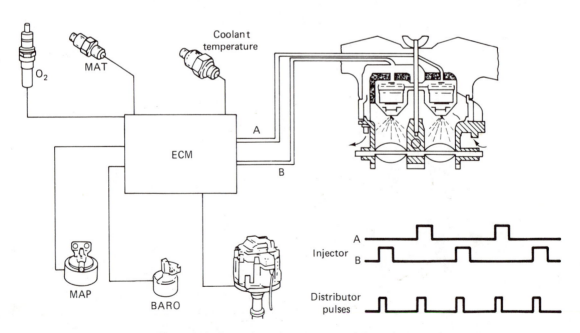

Figure 9-10 Normal running. (Courtesy of General Motors Corporation.)

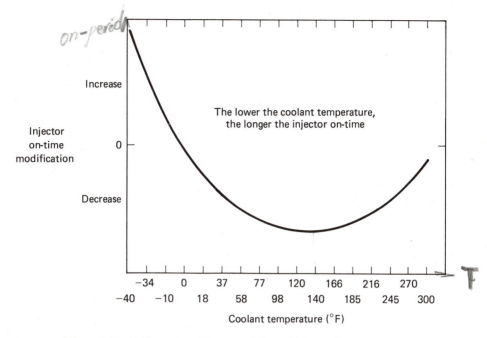

on-period

Injector on-time modification

Increase

0

Decrease

The lower the coolant temperature, the longer the injector on-time

| −34 | 0 | 37 | 77 | 120 | 166 | 216 | 270 |
−40 | −10 | 18 | 58 | 98 | 140 | 185 | 245 | 300

Coolant temperature (°F)

Figure 9-11   Cold running. (Courtesy of General Motors Corporation.)

choke signal in response to changes in manifold vacuum. The timer ensures that the choke does not remain on too long.

Altitude Compensation.   At higher altitudes, fuel does not vaporize as readily when the engine load is increased (and manifold vacuum drops). Unless adjustments are made, the mixture becomes too lean. Consequently, the ECM constantly checks altitude (as indicated by the barometer pressure sensor) and engine load (as indicated by the manifold pressure gauge). When the difference between the two inputs equals a preset PROM value, the program increases the injector on-time and thereby enriches the mixture (Fig. 9-12).

Idle Throttle Compensation.   When the throttle suddenly opens from an idle position, manifold air pressure increases and fuel tends to condense on the walls in the intake manifold. Extra fuel must be added or the mixture will be too lean.

The ECM checks engine rpm and the position of the throttle (Fig. 9-13). If the throttle opens up from an idle position when the engine is below a certain PROM value, longer pulse signals are sent to the injectors. The exact duration of the pulses is influenced by inputs from the manifold air temperature sensor. Since the fuel does not vaporize as well at lower temperatures, the lower the temperature, the longer the pulses.

Acceleration.   Above an idle rpm rate, the ECM regards changes in manifold vacuum as being due to acceleration. However, no matter when or why it occurs, a sudden decrease in manifold vacuum (increase in pressure) causes fuel to condense

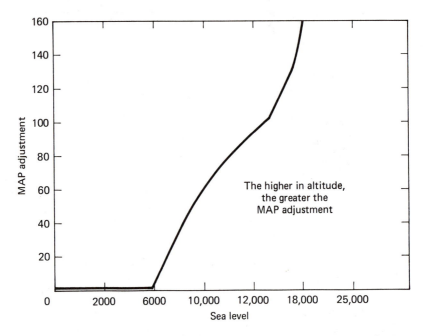

Figure 9-12 Altitude adjustments. (Courtesy of General Motors Corporation.)

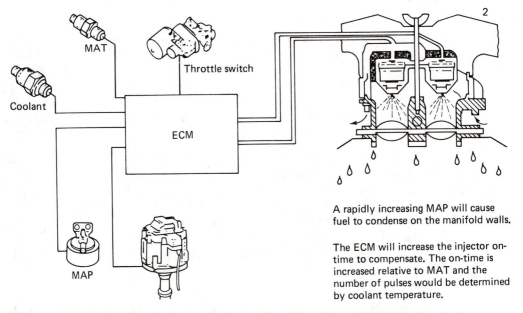

Figure 9-13 Idle speed enrichment. (Courtesy of General Motors Corporation.)

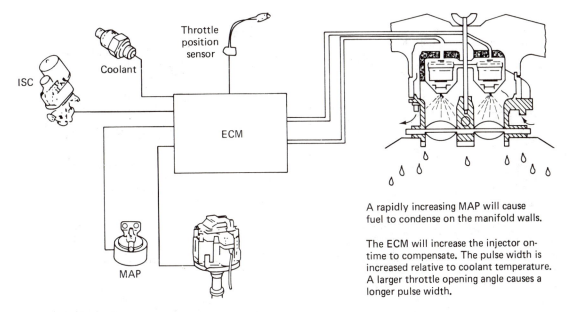

A rapidly increasing MAP will cause fuel to condense on the manifold walls.

The ECM will increase the injector on-time to compensate. The pulse width is increased relative to coolant temperature. A larger throttle opening angle causes a longer pulse width.

Figure 9-14 Acceleration enrichment. (Courtesy of General Motors Corporation.)

and the mixture to become lean. Therefore, manifold air pressure increases above PROM values will result in longer injector pulses (Fig. 9-14). As the pressure decreases, the pulses return to normal duration. To ensure smooth operation, the pulse changes may not occur as fast as the pressure changes. In other words, there may be a lag between the two.

Wide-Open Throttle.   When the difference between barometric pressure and manifold air pressure reaches a certain PROM value, the ECM assumes that the throttle is wide open. As long as this condition remains in effect, the "normal running" inputs are ignored. Injector on-pulses are calculated from the barometric air pressure readings (Fig. 9-15).

Deceleration.   When the throttle plates are closed and the engine slows down, manifold air pressure drops (vacuum increases). This has the opposite effect from the situations previously described, which result in increased pressure (reduced vacuum). Now, any fuel that has condensed in the intake manifold suddenly vaporizes in the presence of the reduced air pressure. The mixture becomes richer. To compensate for this pollution-producing condition, the ECM shortens the injector pulse signals (Fig. 9-16). As a result, less fuel is delivered. However, if the condition occurs within a brief (preset) time of a previous acceleration or idle enrichment operation, the deceleration program will not go into effect. Under those circumstances, the manifold walls are not wet enough to require compensating adjustments.

The ECM looks for the difference
between the MAP and BARO.

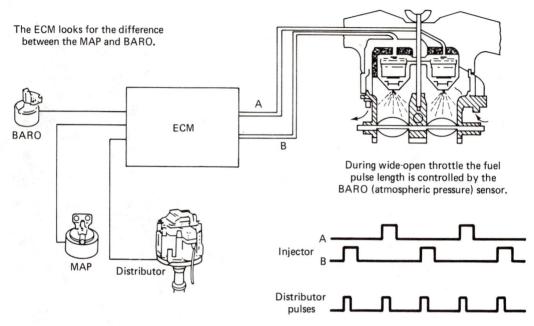

During wide-open throttle the fuel
pulse length is controlled by the
BARO (atmospheric pressure) sensor.

Figure 9-15   Wide-open throttle. (Courtesy of General Motors Corporation.)

Upon deceleration the ECM adjusts
injector on-time based on
battery voltage and closed-loop
fuel control table.

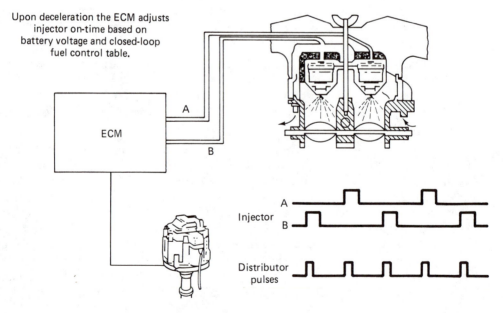

Figure 9-16   Deceleration lean mixture. (Courtesy of General Motors Corporation.)

Function 2: Electronic Spark Timing (EST)

Similar to the control to the injectors, the CCC system also determines the moment of ignition and the coil saturation period (dwell). As in a precomputerized system, the spark is advanced as the engine speeds up. However, instead of relying on centrifugal weights, the basic advance curve is stored in a calibrated PROM (Fig. 9-17).

The timing curve is modified (again, as in a precomputerized system) as manifold pressure fluctuates (Fig. 9-18). The timing curve is also modified in response to inputs from the engine coolant sensor and the barometric pressure sensor. Information from all these sources is converted into certain values based on calibration data contained in a computer storage area called the *look-up tables*. The look-up values are added together to get a combined figure which is then used to determine timing. Of course, these calculations must occur many times a second to ensure that timing adjustments occur in a smooth, inperceptible manner.

In addition to advancing the ignition, controls are provided to retard timing whenever the engine knocks or detonates. The input for this control is a device located on the intake manifold or motor block. The device senses those particular kinds of vibrations which are characteristic by-products of detonation.

The basic reference pulse for the timing operations comes from a pickup coil located on the end of the crankshaft. Using a core and pole piece (similar to the core and pole assembly of the solid-state distributors covered in Chapter 7), the unit pro-

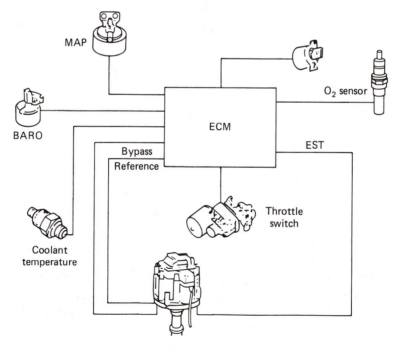

Figure 9-17 Spark timing components. (Courtesy of General Motors Corporation.)

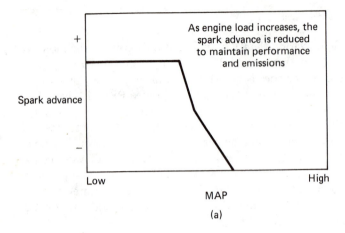

(a)

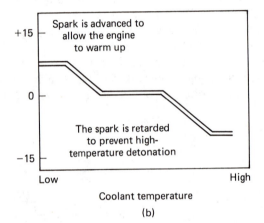

(b)

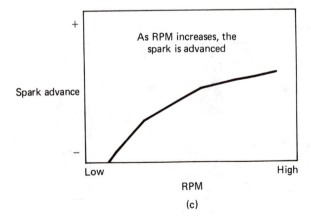

(c)

Figure 9-18   Three conditions that
modify spark timing: (a) MAP modification; (b) temperature modification;
(c) RPM spark modification. (Courtesy
of General Motors Corporation.)

duces a fluctuating (analog) voltage signal (Fig. 9-19). This signal, once converted into digital form, is used to determine engine rpm and piston position. (*Note:* Some GM systems, starting in 1982, use a Hall effect trigger.)

The only times that the crankshaft position does not determine the reference signal to the computer for spark timing are when the engine is being cranked or when the timing is being set. During cranking, a bypass line from the ECM to the distributor drops to 0 volts (Fig. 9-20). This signals the electronic distributor module (the HEI or High-Energy Ignition) to ignore pulses coming from the computer and rely instead on signals generated within the ignition system itself for base ignition timing. When setting or checking the basic timing, the reference signal is disconnected.

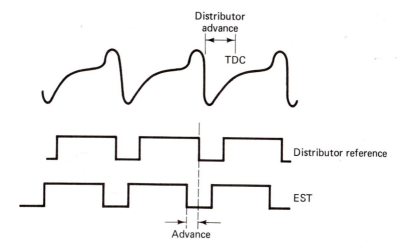

Figure 9-19   Timing reference pulse. (Courtesy of General Motors Corporation.)

Function 3: Idle Speed Control

The main purpose of this program is to control the engine speed during closed throttle operation (Fig. 9-21). The factors that affect idle speed include:

1. *Engine temperature.* When the engine is cold, idle speed is adjusted to approximately 1200 rpm. Then, as the engine warms up, the idle speed is gradually decreased, to about 450 rpm. If the engine temperature exceeds a preset PROM value, the ECM assumes that the engine is about to overheat. It raises idle speed to increase the coolant flow and reduce the temperature.

2. *Battery voltage.* If the battery output falls below a certain level, the idle speed will be increased to help in the recharging process.

3. *Transmission selector.* The throttle is opened when the transmission is shifted to drive or reverse and closed when the selector is moved to park or neutral.

4. *Air-conditioning compressor.* The throttle is also opened slightly when the air-conditioning compressor is engaged. The controls described in items 3 and

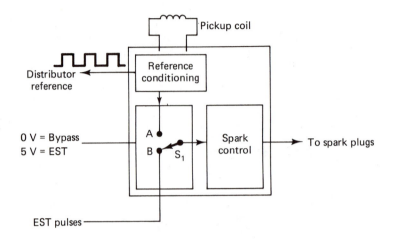

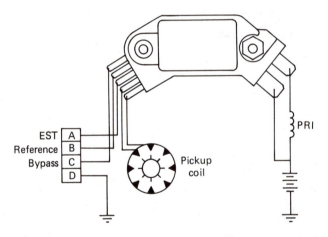

Figure 9-20 Bypass controls. (Courtesy of General Motors Corporation.)

4 are primarily to even out idle speed fluctuations rather than raise or lower idle speed.

The output of the idle-speed control program is a fluctuating voltage signal sent to a motor-driven worm and gear assembly connected to the throttle plates (Fig. 9-22). The output assembly is described in more detail in Chapter 11.

The output device (ISC) is also used to adjust the throttle position during periods of deceleration. This aspect of the idle speed program responds solely to manifold pressure during a certain rpm range. As the manifold pressure drops, the throttle

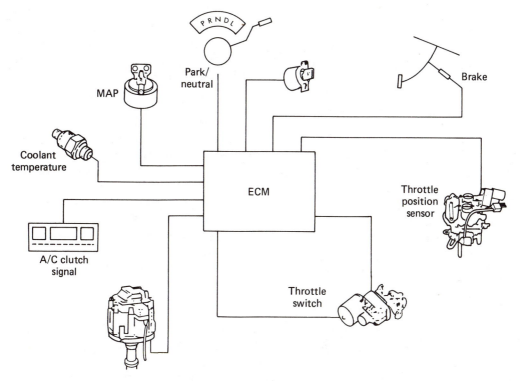

Figure 9-21 Idle speed control. (Courtesy of General Motors Corporation.)

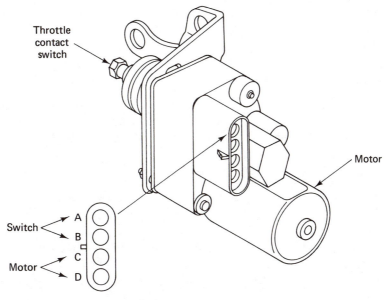

Figure 9-22 Throttle control meter. (Courtesy of General Motors Corporation.)

plate is opened slightly. Opening the throttle plates reduces manifold vacuum, which decreases vaporization and keeps the mixture from becoming excessively rich. Consequently, pollution is reduced. However, opening the throttle also reduces engine braking. So if the brake pedal is pressed more than 4 seconds, the ECM closes the throttle plates, regardless of inputs from the manifold air pressure sensor.

Function 4: Exhaust Gas Recirculation

As we noted in Chapter 3, the EGR valve supplies small amounts of exhaust gas to the intake system whenever manifold vacuum is high enough (Fig. 9-23). The essentially inert exhaust gas does not burn, thereby reducing the combustion chamber temperature and NOx formation.

In GM's computer-controlled system, the EGR passage to the intake manifold is controlled by a solenoid-operated valve. During engine starting and warm-up, a blocking signal is sent from the ECM to the solenoid to close the valve. As a result, exhaust gas recirculation does not take place. However, at all other times, the valve is open, allowing the exhaust gas to be recirculated in response to manifold vacuum.

Function 5: Canister Purge

As in precomputerized systems, a charcoal canister is used to store excess vapors from the fuel tank and fuel system. In the GM system, a solenoid valve controls the purging operation (Fig. 9-24). When the valve is open, trapped gases can be pulled from the canister to the lower-pressure region inside the intake manifold. The solenoid is energized (opened) whenever the engine is in closed loop (normal running) and the coolant temperature is above 80 °C. These conditions are best for burning the excess vapors without producing excessively rich mixtures and thereby increasing pollution.

Function 6: Cruise Control

The cruise control programs manage engine speed. Inputs come from an engine speed sensor, the cruise control on–off switch, a set–coast switch, and a resume–acceleration switch. Outputs go to an accelerator linkage control mechanism.

Function 7: Air Management System

The ECM also controls the flow of air from the air pump. Depending on the input signals, air is directed to the air cleaner, exhaust ports, or catalytic reactor (Fig. 9-25). Two major sets of valves are employed. The first, called the *diverter valve*, sends air to the air cleaner or to the second valve, which is called the *switching valve*. This second valve sends air to the exhaust ports or to the catalytic reactor. The exhaust air passages also contain the usual check valves to prevent backfiring during deceleration.

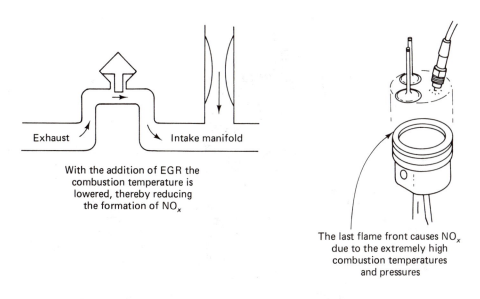

With the addition of EGR the combustion temperature is lowered, thereby reducing the formation of NO_x

The last flame front causes NO_x due to the extremely high combustion temperatures and pressures

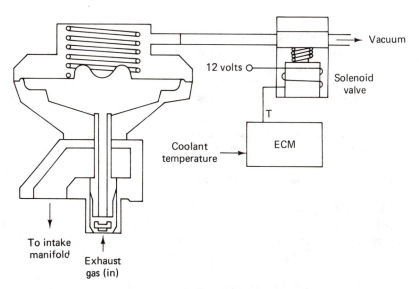

Figure 9-23 EGR control. (Courtesy of General Motors Corporation.)

The system has three main modes of operation:

1. During normal or closed-loop operation, the diverter valve sends air to the switching valve, which, in turn, sends air to the reactor. The catalytic agents in the reactor are formed in two layers or beds (Fig. 9-26). Air injected between the beds helps the platinum and palladium agents oxidize HC and CO.

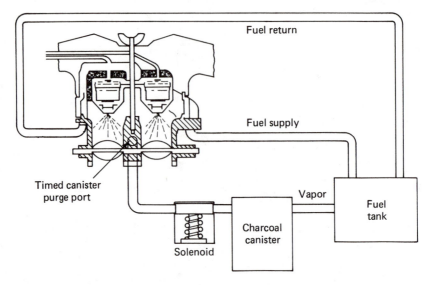

Figure 9-24 Canister purge. (Courtesy of General Motors Corporation.)

2. During cold operation and in other open-loop conditions, the reactor is not hot enough to make use of the extra air. Consequently, the diverter valve sends air to the switching valve, which directs the air to the exhaust ports. Performing the function normally assigned to pressurized air from the air pump (until computerized systems), it helps oxidize HC and CO contained in the exhaust gases.

3. When the reactor gets too hot, air added to the exhaust or the reactor can cause damage. Under those conditions, the diverter valve sends air to the air cleaner.

Function 8: Torque Convertor Clutch

Still another application for the on-board computer is controlling an automatic transmission clutch (Fig. 9-27). Normally, there is a certain amount of slippage or wasted energy in a torque convertor or automatic transmission. As the transmission turbine forces fluid through the stator to the pump, the turbine has a tendency to rotate faster than the pump.

The speed difference between the engine and the transmission can be eliminated by a direct connect clutch, pressure plate, spring damper, and control solenoid assembly. The solenoid, which determines the position of the clutch, is energized by the ECM in response to inputs from these sources:

1. *Brake switch:* If the brake pedal is pressed, the solenoid will be energized regardless of inputs from the other sources. The transmission will operate in a normal manner.

2. *Transmission pressure switch:* This switch tells the ECM what gear the transmis-

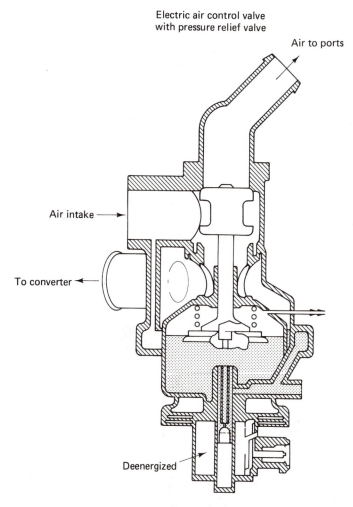

Figure 9-25 Air management components. (Courtesy of General Motors Corporation.)

sion is in. Different vehicles use the clutch feature for different gears. The exact combination is stored as one of the PROM calibration values.

3. *Park-neutral switch:* The control solenoid is disengaged when the transmission is in park or neutral.

4. *Vehicle speed sensor:* The solenoid is energized only after the vehicle reaches a certain preset PROM value (usually 35 to 45 mph).

5. *Coolant temperature sensor:* The solenoid is also engaged only after the engine reaches a certain preset temperature.

6. *Manifold air pressure sensor:* When the engine is operating above a certain preset load (as indicated by the manifold pressure reading), the clutch is released.

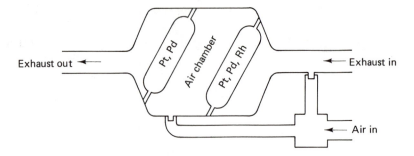

Figure 9-26   Dual-bed convertor. (Courtesy of General Motors Corporation.)

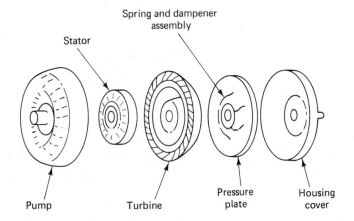

Figure 9-27   Torque convertor clutch. (Courtesy of General Motors Corporation.)

Function 9: Early Fuel Evaporation

Responding to inputs from the temperature sensor, the computer also controls the operation of the exhaust heat riser valve. As the engine warms up, the position of the valve is adjusted for the best combination of emission control and vehicle performance.

Function 10: Modulated Displacement Engine

One of the most interesting aspects of the CCC system is the modulated displacement option on large Cadillac V-8 engines. Designed to improve fuel economy, the system responds to inputs from the speed sensor, transmission gear sensor, and the engine coolant sensor. When the correct combination of inputs is encountered, the computer sends out signals to stop the operation of two or four cylinders. It does this by directing control solenoids to raise the pivot point of the appropriate intake and exhaust valves so that the valves no longer open (explained in more detail in

Chapter 11). Responding to different signals from the normal input sources, fuel metering and ignition timing are also changed.

Here are some of the control factors affecting engine displacement modulation.

1. Operation in six- or four-cylinder modes occurs only when:
   (a) Coolant temperature is above 120°F.
   (b) The engine is in closed-loop operation.
   (c) The transmission is in third gear.
2. Transition from eight- to six- or four-cylinder operation is not allowed at engine speeds over 2600 rpm. That is because there is not enough time between valve opening and closing for the control solenoids to operate.
3. Shifts from eight- to four-cylinder operation can occur at speeds above 24 mph.
4. Shifts from eight- to six-cylinder operation, however, cannot occur at speeds below 47 mph. That is because six-cylinder operation causes uneven firing impulses and rough running. These conditions are pronounced at lower engine speeds, but not very noticeable at higher speeds.

Function 11: Instrument Panel Display

This is called the Driver Information System by GM. Using a dashboard display panel, the system provides the following information:

1. Average miles per gallon
2. Instantaneous miles per gallon
3. Range (distance) with the remaining fuel in the tank
4. Number of cylinders in operation (modulated displacement engines only)

The particular information displayed depends on certain selection buttons pressed by the driver. By pressing the required buttons, the driver can cause the desired information to be displayed.

GM uses a separate computer to calculate driver information. However, it is possible to use the same computer to manage the engine and display driver-oriented facts. *Note:* See the end of this chapter for a more complete discussion of digital instruments.

Function 12: System Self-Diagnosis

The ECM constantly checks certain input and output information to make sure that it is within preset limits. The basic categories of information examined include:

1. Switch inputs to the ECM, to determine if certain devices are on or off at the proper times
2. Analog sensor inputs, to make sure that the fluctuating voltage produced by particular sensors is within predefined limits

3. Switch output signals, to make sure that the correct devices are being turned on or off

If the ECM detects a failure, a failure code is stored in the RAM memory and "CHECK ENGINE LAMP" on the dashboard is illuminated. This alerts the driver to a possible problem. If the failure is intermittent, the light goes out as soon as the problem corrects itself. However, the failure code remains in memory.

Failure codes are displayed on a dashboard readout panel by pressing a certain sequence of buttons on the air conditioning control unit. First a confirmation code appears, telling the mechanic that the selection has been made properly; then the failure codes appear in numeric order. (*Note:* On some GM cars, the correct terminals on a test connector must be jumped to obtain troubleshooting codes.) Details of these and other troubleshooting systems are covered in detail in Chapter 13.

Fail-Safe Operation

Because the computer is a vital part of engine operation, it constantly checks itself to make sure that the programs are working properly. On encountering certain problems, a backup injector program is put into action. The backup program, which uses fewer inputs, and consequently is less sensitive to sensor malfunctions, provides three injector pulse durations:

1. 15 milliseconds for cranking
2. 2 milliseconds for idle
3. 5 milliseconds for open-throttle operation

These pulse durations will not give the best operation for all conditions, but will allow the vehicle to be driven to the garage for repairs.

In addition to the fuel injection backup program, the system also provides an ignition system backup. In the event of a problem, the bypass line noted previously drops to 0 volts. This tells the ignition system to ignore inputs from the ECM and to rely instead on signals produced within the ignition system.

CHRYSLER CONTROL SYSTEMS

In 1974, Chrysler introduced one of the first computerized control systems. Called the Lean Burn system, it controlled ignition timing with an electronic analog computer. This electronic analog unit, like mechanical analog devices, differs from digital computers in the way it represents information. Instead of storing information as 1's and 0's, it uses continuously variable voltages as an "analog" or representation of information to be processed. The programming is essentially fixed and cannot be changed except by changing the physical design of the unit.

As of this writing, Chrysler still uses an analog computer (the ESA or Electronic Spark Advance) to control ignition timing. However, like other companies,

Chrysler has also added digital electronic controls for additional engine operations. Termed the Combustion Control Computer (CCC) (Fig. 9-28), the system has four separate circuits (including the ESA). They are:

1. Electronic Fuel-Injection (EFI) circuit for regulating the speed of a fuel-injection pump and thereby controlling the air/fuel mixture
2. Auto Calibration Circuit for making fine-tuning adjustments to the EFI circuit
3. Electronic Spark Advance (ESA) for advancing or retarding the spark
4. Automatic Idle Speed (AIS) circuit for adjusting the idle speed

Inputs

These sensors and sensing functions supply information to the control circuits:

1. Airflow sensor
2. Fuel flowmeter
3. Fuel temperature sensor
4. Fuel pressure switch
5. Throttle position potentiometer
6. Closed throttle switch
7. Coolant temperature sensor
8. O_2 sensor
9. Ignition switch sensor
10. Ignition pulse sensor
11. Air conditioning on–off switch
12. Engine speed sensing (calculating from ignition pulses)
13. Engine load sensing (calculated from engine speed and airflow rate)

Output Devices

Acting on information from the input sources listed above, the computer control circuits adjust the operation of these components:

1. Fuel injection pump
2. Ignition coil primary circuit
3. Dc-powered throttle position motor

Control Function

The functions performed by the Chrysler system are similar, in purpose at least, to those provided by the GM system. The primary differences are in the number of functions performed—the GM system does more—and in the number of mechanical, pneumatic, and hydraulic analog devices blended into the Chrysler system. The fol-

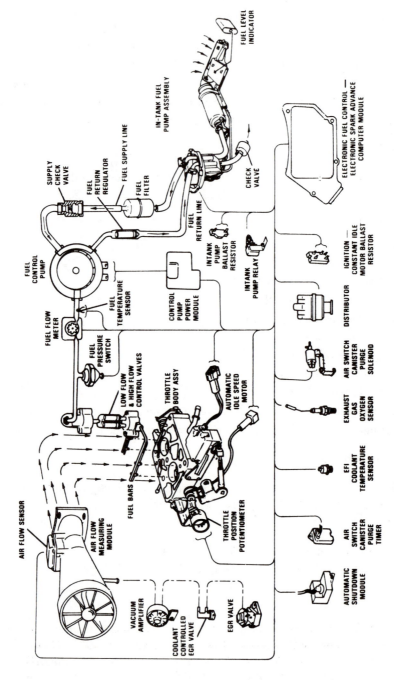

Figure 9-28 CCC components. (Courtesy of Service and Parts Division, Chrysler Corporation.)

FUEL LEVEL INDICATOR

IN-TANK FUEL PUMP ASSEMBLY

FUEL SUPPLY LINE

FUEL RETURN REGULATOR

SUPPLY CHECK VALVE

FUEL FILTER

CHECK VALVE

FUEL RETURN LINE

ELECTRONIC FUEL CONTROL — ELECTRONIC SPARK ADVANCE COMPUTER MODULE

FUEL CONTROL PUMP

FUEL FLOW METER

FUEL TEMPERATURE SENSOR

CONTROL PUMP POWER MODULE

INTANK PUMP BALLAST RESISTOR

INTANK PUMP RELAY

IGNITION CONSTANT IDLE MOTOR BALLAST RESISTOR

DISTRIBUTOR

FUEL PRESSURE SWITCH

LOW FLOW & HIGH FLOW CONTROL VALVES

THROTTLE BODY ASSY

AUTOMATIC IDLE SPEED MOTOR

AIR SWITCH CANISTER PURGE SOLENOID

EXHAUST GAS OXYGEN SENSOR

AIR FLOW SENSOR

AIR FLOW MEASURING MODULE

FUEL BARS

THROTTLE POSITION POTENTIOMETER

EFI COOLANT TEMPERATURE SENSOR

AIR SWITCH CANISTER PURGE TIMER

VACUUM AMPLIFIER

COOLANT CONTROLLED EGR VALVE

EGR VALVE

AUTOMATIC SHUTDOWN MODULE

lowing paragraphs discuss the main functions performed by each of the four Chrysler electronic control circuits.

Electronic Fuel-Injection (EFI) Functions

The primary purpose of this subsystem is controlling the operation of a fuel-injection pump (Fig. 9-29). The pump is located over a throttle body, which, in turn, is positioned on top of the intake manifold.

The injection pump, or *control pump* as it is called, receives fuel under pressure from a delivery pump located in the fuel tank (Fig. 9-30). The delivery pump maintains a constant supply of fuel to the control pump. A pressure regulator (analog control device) sends excess fuel back through a return line to the fuel tank.

Among the mechanical devices included in the fuel control system are the fuel

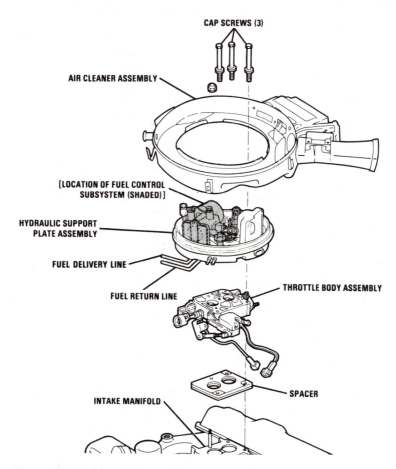

Figure 9-29   Fuel control system. (Courtesy of Service and Parts Division, Chrysler Corporation.)

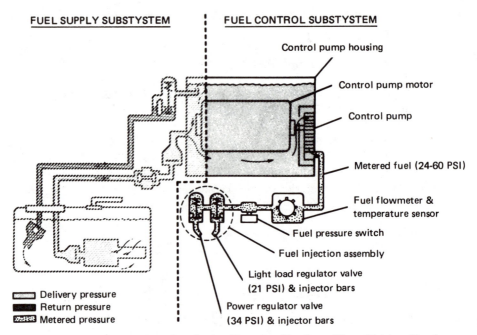

Figure 9-30 Schematic of fuel flow circuits. (Courtesy of Service and Parts Division, Chrysler Corporation.)

control pump, a fuel flowmeter, a fuel temperature sensor, a fuel pressure switch and two fuel injector bars, and regulator valves (Fig. 9-31). The injection bars are U-shaped devices positioned over the throttle body air intake. Fuel sprays out of tiny holes on the underside of the bars. One bar is designated for light-load operation and is supplied by a control valve, which opens when the fuel pressure exceeds 21 pounds per square inch (psi). The control valve for the other injection bar opens up when the fuel pressure exceeds 34 psi. It is called the *power bar*.

Although different from the GM system mechanically, the Chrysler fuel injection system also provides control over several distinct operating ranges. They include:

1. Initial startup
2. Open-loop operation
3. Closed-loop operation

Initial startup.   Immediately after the ignition switch is turned to the START position, the fuel-injection pump comes under the control of the fuel pressure switch. Sensing little or no pressure, it signals the control pump to operate at full speed. This rids the fuel flow circuit of air and primes the fuel flow sensing device.

As soon as the fuel pressure builds up sufficiently to be registered by the fuel flow sensor, control is transferred to the EFI computer circuit. As in the GM system, there are two basic modes of operation, open and closed loop.

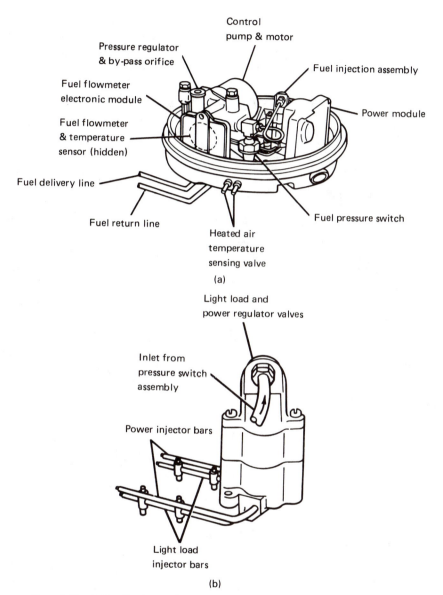

Figure 9-31   (a) Details of control purge assembly; (b) fuel injection assembly. (Courtesy of Service and Parts Division, Chrysler Corporation.)

Open-loop operation.   The system is considered to be in open-loop operation any time the O_2 sensor is not providing the primary control input for the fuel injection pump (Fig. 9-32). The major periods of open-loop operation are when the engine is too cold for the O_2 sensor to be operative and when fuel enrichment is required for increased load or cold-operating conditions. During open-loop operation, the primary inputs come from the coolant temperature sensor and the engine speed and load sensing devices. The primary output source is the fuel injection control pump. It is speeded up to provide more fuel for choking and engine loading. As the inputs change, the pump gradually returns to a slower speed.

Closed-loop operation.   The system operates in a closed-loop manner whenever the O_2 sensor is at its normal operating temperature and the engine is running more or less at a steady speed (Fig. 9-33). As in the GM system, signals from the O_2 sensor determine the operation of the injection pump, which in turn affects the readings.

Auto Calibration Circuit

Operating as a separate but related computing device, this circuit is used to monitor and adjust the control provided by the EFI circuit. Examining voltage signals from the O_2 sensor, as well as other inputs not available to the EFI, the calibrating circuit determines what levels of input relate to the stoichiometric ratio. During steady speeds (around 55 mph) and at idle, it compares its own assessment of the stoichiometric ratio with the assessment maintained in the EFI circuit. If different, the EFI is recali-

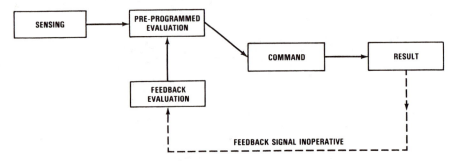

Figure 9-32   Open-loop operation. (Courtesy of Service and Parts Division, Chrysler Corporation.)

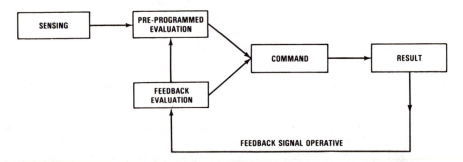

Figure 9-33   Closed-loop operation. (Courtesy of Service and Parts Division, Chrysler Corporation.)

brated to correspond to the calibration circuit. This function will adjust for any factors that affect EFI calibration. However, the primary purpose is to compensate for barometric changes.

Electronic Spark Advance (ESA) Circuit

This analog electronic computer advances and retards the spark in response to input from an engine speed sensor and an airflow meter. The engine speed sensor initiates spark advance as the engine speeds up. The airflow rate going into the air cleaner and engine speed are examined to determine engine load. As in all other systems, after a given load, the spark is retarded as the load increases.

Automatic Idle Speed (AIS) Circuit

The programs running in this circuit make sure that the idle speed remains above certain predefined levels. Input information is examined regarding (1) the engine coolant temperature, (2) the engine speed, (3) the throttle position, and (4) the air conditioning pump. The principle output is to a variable-speed motor connected to the throttle plates. Throttle plate adjustment is divided into three main levels of control:

Cold engine idle speed.   The colder the engine, the wider the throttle opening and the faster the idle speed. As the engine warms up, the idle speed gradually falls off to a predefined minimum. This minimum is considered to be the normal operating temperature idle speed.

Cranking idle speed.   When the engine is first cranked the AIS examines inputs regarding throttle position, coolant temperature, and engine speed. If the throttle is closed, the AIS will direct the control motor to open the throttle to the limit defined by the temperature reading. Then, as the engine starts to run on its own and the idle speed exceeds the predefined minimum, the throttle starts to close. It will continue to close until the idle speed reaches the programmed limit to the particular temperature.

Air-conditioner pump.   When the air-conditioner compressor comes on, the AIS will maintain the previous idle speed in spite of the increased load on the engine.

Fail-Safe Features

Like the GM system, the Chrysler CCC circuits maintain backup programs which are enacted when problems occur. The following contingencies are covered:

1. The computer system constantly checks for signals from the ignition system. During normal operation, these signals help time the operation of other functions. However, if the ignition signal drops below a certain level (correspond-

ing to an idle speed of 150 rpm), and the ignition key is in the RUN or ON positions, power to the EFI circuit will be interrupted.

2. During cranking, power is maintained to the EFI circuit even if no spark signal is present. The EFI is shut down only if the control pump is signaled to operate at full speed for more than 10 to 20 seconds. Then the key has to be turned all the way back to the OFF position before power can be restored to the EFI.

BOSCH MOTRONIC SYSTEM

The Bosch Motronic system is manufactured by the Robert Bosch Company of Germany. A fully digital unit, it provides most of the functions performed by other systems. The on-board computer manages fuel injection and ignition timing during both normal (closed-loop) running and open-loop operation (acceleration, cranking, cold operation, etc.). Figure 9-34 shows the major components.

However, the Bosch system is different in two respects. One, as shown in the illustration, is the fuel-injection technique. Instead of using throttle body injection, as is common in many other systems, the Bosch unit pictured here employs multiple fuel injectors. One injector is located near the intake port of each cylinder.

These injectors operate in unison. In other words, every time the computer sends an injection pulse, all the injectors squirt fuel at the same time. (This is also different from similar fuel injection systems used in precomputerized engines. Then, each injector operated independently and was timed to supply fuel on the intake stroke of the individual cylinders.) As in other computerized injection systems, the Bosch Motronic injection pulses are modulated by the computer in response to signals received from the input devices.

The second principal difference between the Bosch system and other systems relates to the way the Bosch unit is sold. Bosch does not produce engines; it sells equipment to manufacturers who do. Therefore, the Bosch computer control system could be found on any engine, especially those produced by European manufacturers. For this reason it is likely that the Motronic hardware and software (firmware) would be modified by the end user.

DIGITAL INSTRUMENTS

A number of manufacturers now supply digital instruments on many of their cars. First introduced as extra-cost options, these systems may become a standard equipment before long.

For the driver, one major difference between digital systems and heretofore conventional systems is the way in which information is displayed on the dash. Conventional instruments use an analog display; that is, the reading is directly "analogous" to the condition being measured. Dial-type speedometers, clocks, and fuel gauges are examples of analog information displays.

Digital instruments, on the other hand, generally convert analog information

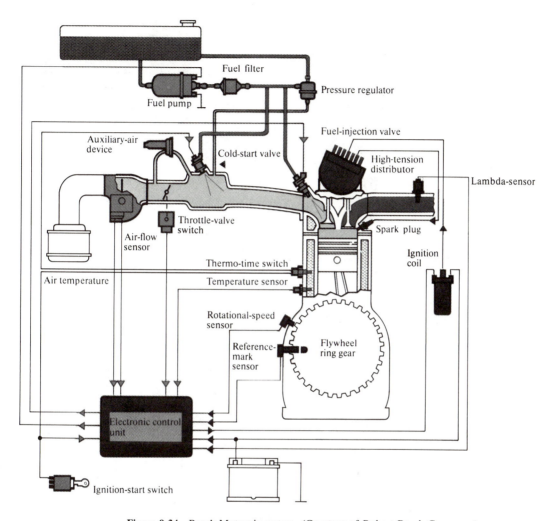

Figure 9-34   Bosch Motronic system. (Courtesy of Robert Bosch Company.)

directly into numbers or symbols. For instance, rather than displaying speed as the position of a pointer on a dial face, a digital speedometer shows speeds as a particular number: 55 mph, 60 mph, and so on. (NOTE: Some digital systems display information in an analog fashion using bar charts and visuals that resemble dial type instruments. However, the processing is still done digitally.)

Another difference between digital and analog systems is the amount of "calculated" information available. Although analog devices can perform information processing, as we noted earlier in the book, such devices are larger and less flexible than digital computers. Therefore, analog dash instruments have been limited to supplying basic information. Even when a dashboard includes such sophisticated instruments as vacuum gauges, voltmeters, ammeters, oil pressure gauges and tachometers no attempt has been made to automatically determine correlations in the data. That is up to the driver.

However, when this information is converted to digital form and presented to a digital computer, calculations are possible. Programs can be devised for the on-board engine computer or for a separate dashboard computer. Depending on the amount of input data available, these programs can perform various services for the driver. At the present time, most programs are used to calculate trip-related data, such as fuel consumption, distance to and estimated arrival time at a given destination, and so on. However, later, more sophisticated programs will probably be developed.

To get an idea of the kinds of digital instruments being used, the following paragraphs briefly examine the system used on Mark VI Lincolns manufactured by Ford Motor Company.

FORD ELECTRONIC INSTRUMENT SYSTEMS

The main components in this system (Fig. 9-35) include:

1. Electronic fuel gauge
2. Electronic speedometer
3. Message center display
4. Selection keyboard
5. A small computer, mounted behind the dashboard, for the message display and the keyboard
6. Various sensors located around the engine and vehicle

These components are used to provide three basic categories of information:

1. Ongoing driver information
2. Warnings
3. Processed data

Ongoing Driver Information

These data include the information appearing on the electronic fuel gauge and the electronic speedometer. Both these instruments are connected to more-or-less conventional sensors. The fuel gauge receives data from a standard variable resistor. The only difference is that high resistance indicates a full tank, low resistance a low tank—just the opposite of previous systems. The electronic speedometer is connected to a mechanical cable. The major difference is in the head. It turns an optical light sensor, which sends digital pulses to the display unit.

In both these systems, the primary information processing is handled by mechanical or electromechanical sensors. Digital processing occurs only in the conversion from analog to digital symbols.

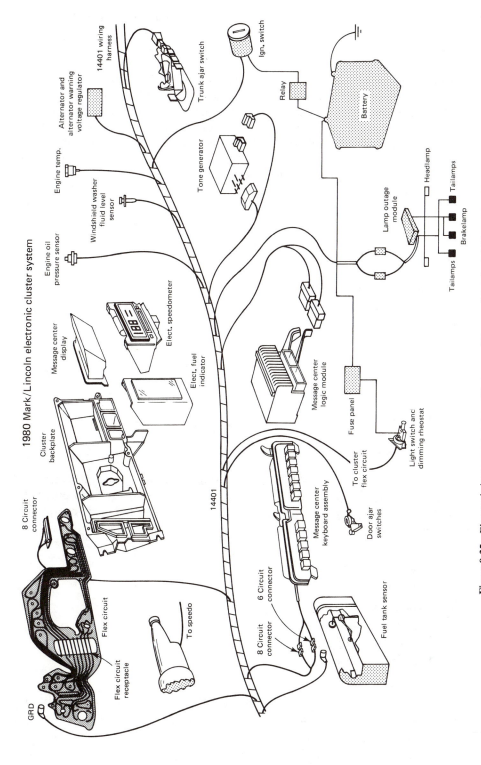

Figure 9-35 Electronic instrument system. (Courtesy of Parts and Service Division, Ford Motor Company.)

1980 Mark/Lincoln electronic cluster system

14401 wiring harness

Alternator and alternator warning voltage regulator

Trunk ajar switch

Ign. switch

Relay

Battery

Engine temp.

Tone generator

Windshield washer fluid level sensor

Engine oil pressure sensor

Lamp outage module

Headlamp

Tailamps

Brakelamp

Tailamps

Message center display

Elect. speedometer

Message center logic module

Fuse panel

Elect. fuel indicator

Cluster backplate

8 Circuit connector

14401

Message center keyboard assembly

To cluster flex circuit

Light switch and dimming rheostat

Door ajar switches

Flex circuit

Flex circuit receptacle

To speedo

8 Circuit connector

6 Circuit connector

Fuel tank sensor

GRD

Warnings

Warning information is displayed as printed words on the message center display panel. These messages appear when sensors located around the vehicle and engine detect certain conditions. The conditions examined are mostly of an either–or, on–off nature: fuel level below a certain value, door or trunk ajar, oil pressure below a certain value, and so on. Therefore, the sensors are mainly switches, opening or closing electrical signal circuits. Primary information processing occurs when the message display converts this inherently digital data into printed words.

The warnings included in the Ford system include:

1. Low fuel
2. Alternator
3. Low brake pressure
4. Low washer fluid
5. Headlamp out
6. Taillamp out
7. Brake lamp out
8. Truck ajar
9. Door ajar
10. Oil pressure low
11. Engine temperature high

Processed Data

The message center is also used to display information processed by the small digital computer located behind the dashboard (Fig. 9-36). By pressing the appropriate key on the dash-mounted keyboard (Fig. 9-37), the driver can obtain this trip-related information:

1. Distance traveled
2. Elapsed time
3. Average speed
4. Distance to destination
5. Estimated time of arrival
6. Distance to empty (fuel)

The computer determines these facts based on information supplied by the electronic speedometer, the on-board clock, and certain data provided by the driver.

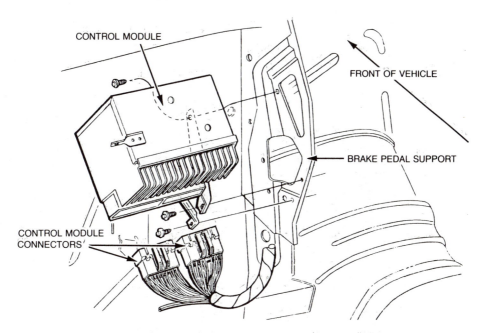

Figure 9-36 Dash-mounted instrument computer. (Courtesy of Parts and Service Division, Ford Motor Company.)

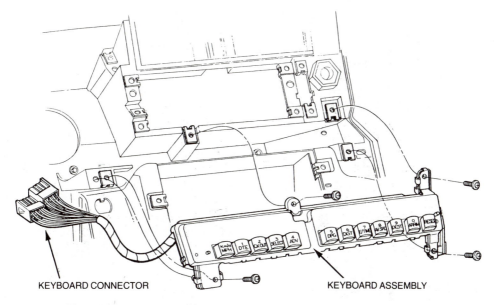

Figure 9-37 Dash-mounted keyboard. (Courtesy of Parts and Service Division, Ford Motor Company.)

10

Input Devices

This chapter describes the operation of the input devices that supply information to on-board computers. Since the units are similar from one manufacturer to another, the discussion will center on selected sensors and signal devices used in the Ford 1980 EEC-III system and the 1981 GM CCC system.

REMINDER (DIGITAL VERSUS ANALOG OPERATION)

Before getting into the discussion of specific input devices, it is important to note that each one of these sensors converts a physical condition or state into an electrical signal that represents the condition. Some devices just produce an on–off or two-state output. Such information is essentially digital and does not need to be converted. However, in many cases, including most of those discussed in this chapter, the sensor output (or computer input) is a continuously variable voltage. These signals are "analogs" or representations of the physical condition being measured and must be converted by the computer into a digital pattern of 1's and 0's.

REFERENCE VOLTAGES

Another point to bear in mind is that most of these sensing units employ reference voltages. A reference voltage is a known voltage signal sent out by the computer to the sensor. The sensor modifies the reference signal, usually by passing it through a variable resistor network. The resistance at any moment depends on the state of the physical condition being measured. After being modified, the reference signal is returned to the computer. By comparing the altered return signal with the known reference voltage, the computer is able to assign a numeric value to the physical condition.

GM PRESSURE SENSORS

GM employs two basic types of air pressure sensors, absolute and differential (Fig. 10-1). *Absolute units* compare manifold or barometric (atmospheric) pressure to a

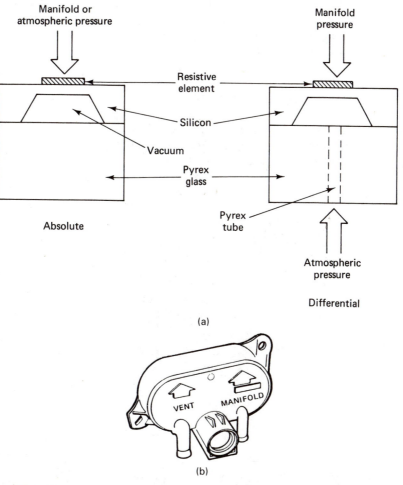

(a)

(b)

Figure 10-1 (a) GM pressure sensors (Courtesy of General Motors Corporation); (b) Ford pressure sensor (Courtesy of Parts and Service Division, Ford Motor Company).

fixed reference pressure. Such devices can be used to determine altitude by comparing the current barometric reading to a sea-level reference pressure. *Differential units* directly compare manifold and barometric pressures. They are used to calculate manifold pressure readings (corrected for altitude).

Both types of units contain flexible, metalized quartz plates. The plates bend in response to changing pressures from one or two sources. As the distance between the plates varies, the capacitance of the unit also varies. Consequently, a reference voltage supplied to the assembly will go up or down depending on the pressure(s) involved.

The primary difference between absolute and differential units is the design of the quartz plate sensing unit. In absolute units, both plates surround a sealed reference pressure cavity. The bending action is a product of one variable pressure source and the fixed reference pressure. Differential units open up one side of the chamber so that atmospheric and manifold pressure work against the same sensing plate. The resulting flexing action is a product of both sources.

FORD COOLANT TEMPERATURE SENSOR

This unit consists of a thermistor located inside a brass plug (Fig. 10-2). The entire assembly is threaded into the water heater outlet. As the temperature of the coolant changes, the electrical resistance of the sensor varies in proportion to the temperature. Consequently, the reference voltage supplied to the ECM also varies in proportion to the temperature.

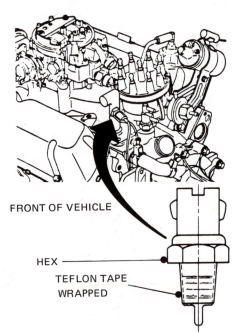

FRONT OF VEHICLE

HEX

TEFLON TAPE
WRAPPED

Figure 10-2 Ford coolant temperature sensor. (Courtesy of Parts and Service Division, Ford Motor Company.)

FORD THROTTLE POSITION SENSOR

This sensor lets the computer know whether the throttle is closed, partially opened, or fully opened. The main sensing element is a variable resistor, or potentiometer, mounted on the throttle shaft (Fig. 10-3). As the shaft turns, the resistance is varied, which alters the reference voltage output to the computer. (The same sort of control

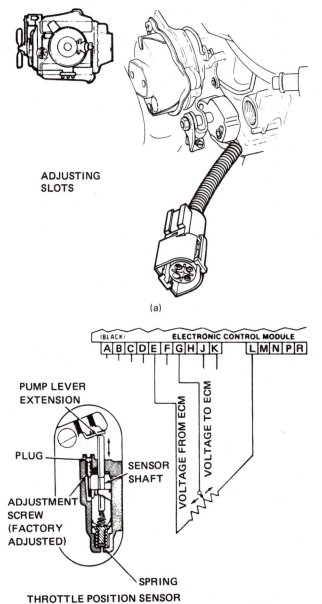

ADJUSTING
SLOTS

(a)

(BLACK)   ELECTRONIC CONTROL MODULE

A B C D E F G H J K   L M N P R

PUMP LEVER
EXTENSION

PLUG

SENSOR
SHAFT

VOLTAGE FROM ECM

VOLTAGE TO ECM

ADJUSTMENT
SCREW
(FACTORY
ADJUSTED)

SPRING

THROTTLE POSITION SENSOR

(b)

Figure 10-3 (a) Ford throttle position sensor (courtesy of Parts and Service Division, Ford Motor Company); (b) GM throttle position sensor (courtesy of Chevrolet Motor Division, General Motors Corporation).

is used to adjust the volume of a TV set or radio.) The entire assembly is attached to the throttle body on a slotted mount. That way the unit can be adjusted back and forth for calibration.

FORD CRANKSHAFT POSITION SENSOR

The crankshaft position sensor works somewhat like the pulse ring (reluctor) and pickup coil in a breakerless ignition system. Producing signals from a pulsating magnetic field, it informs the computer about the position of the crankshaft. This information is then used to determine engine speed and help calculate ignition timing.

The main components of the sensor are a steel pulse ring and pickup coil (Fig. 10-4). The pulse ring is pressed onto the crankshaft damper, which, in turn, is keyed to the front of the crankshaft. The ring has four lobes spaced 90° apart (no more are needed for an eight-cylinder engine). The lobes are positioned to provide a built-in timing advance of 10°. No provisions are made for adjustment, since the unit is fixed at the time of manufacture.

The pickup sensing coil is mounted on the front of the engine and aligned with the pulse ring. As the lobes approach the coil, the sensor output voltage increases; then, as the lobes pass, the voltage diminishes to the base-level reference voltage supplied by the computer. The number of times the voltage changes (e.g., the number of changes from a maximum to minimum voltage) is directly proportional to the engine speed.

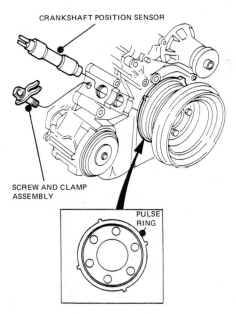

CRANKSHAFT POSITION SENSOR

SCREW AND CLAMP
ASSEMBLY

PULSE
RING

Figure 10-4   Ford crankshaft position sensor. (Courtesy of Parts and Service Division, Ford Motor Company.)

FORD EGR VALVE SENSOR

This sensor monitors the position of the pintle valve used in the EGR (exhaust gas recirculation) system. The valve itself is operated by a vacuum diaphragm and spring assembly (Fig. 10-5). Reduced manifold pressure causes the diaphragm to open the pintle, thereby allowing exhaust gas to mix with the incoming air and fuel. When manifold pressure increases, the return spring closes the valve.

The sensor, which is mounted on the diaphragm, contains a variable resistor. As the diaphragm moves up and down, the resistor modifies a reference signal from the computer. That way, the computer is able to judge the position of the pintle valve.

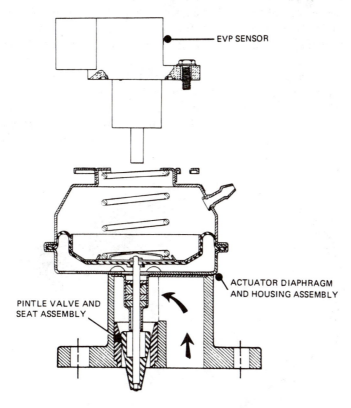

Figure 10-5   Ford EGR pintle position sensor. (Courtesy of Parts and Service Division, Ford Motor Company.)

The sensor itself does not determine the position of the valve. That control comes from another valve, which, on command from the computer, blocks the vacuum passage to the EGR valve diaphragm. When directed to open, the valve lets reduced pressure from the intake manifold draw in exhaust gas.

GM OXYGEN SENSOR

The O_2 sensor (Fig. 10-6) is actually a small battery. The main components are two platinum plates with a zirconia electrolyte located in between. Positioned in the exhaust manifold, the sensor directs hot exhaust gas to one plate and outside air to the other (Fig. 10-7). Here is how the unit works:

When the zirconia electrolyte is exposed to oxygen, it becomes a carrier of free electrons. The platinum plate nearest the outside air is exposed to more oxygen, so it has more electrons. It becomes negatively charged. The plate nearest the exhaust gas encounters fewer atoms of oxygen, giving it a positive charge. If the two plates are connected into a complete circuit, current will flow.

The voltage or pressure causing the current flow depends on the potential between the two platinum plates. It can vary from 100 to 900 millivolts. The exact reading depends on the O_2 content in the exhaust, which, in turn, depends on the air/fuel ratio.

One essential aspect of the sensor is the speed at which it reacts to changes in the air/fuel ratio. As noted in Fig. 10-8, the voltage curve changes abruptly when the fuel mixture reaches the stoichiometric ratio of 14.7:1. Leaner mixtures produce a voltage level near 900 mV; richer mixtures produce a voltage level near 100 mV. The computer examines the voltage fluctuation, then makes corresponding changes in the injector on-time.

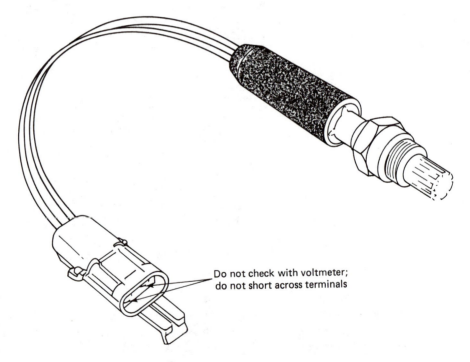

Do not check with voltmeter; do not short across terminals

Figure 10-6   GM O_2 sensor. (Courtesy of General Motors Corporation.)

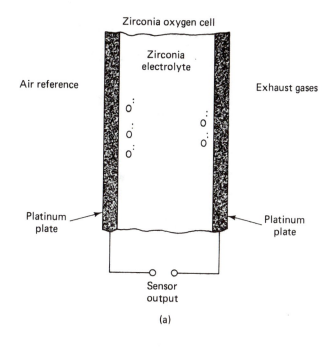

Zirconia oxygen cell

Zirconia electrolyte

Air reference

Exhaust gases

Platinum plate

Platinum plate

Sensor output

(a)

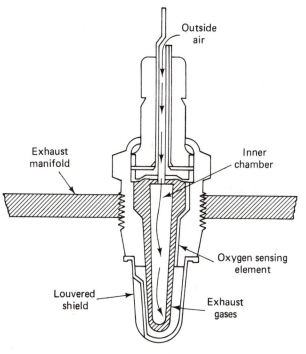

Outside air

Exhaust manifold

Inner chamber

Oxygen sensing element

Louvered shield

Exhaust gases

(b)

Figure 10-7 Operation of O_2 sensor. (Courtesy of General Motors Corporation.)

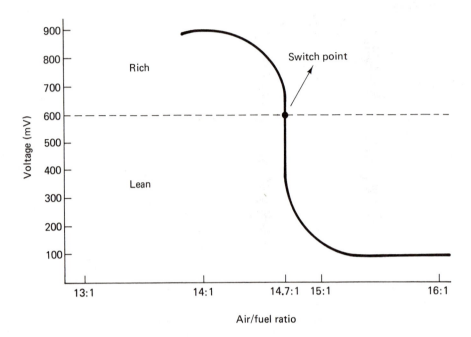

Figure 10-8 Sensor voltage curve. (Courtesy of General Motors Corporation.)

However, as we saw in Chapter 9, this closed-loop method of operation works only when the exhaust is above certain temperature levels. At levels below 200°C, the voltage does not switch fast enough to provide useful readings. Figure 10-9 shows a Ford O_2 sensor.

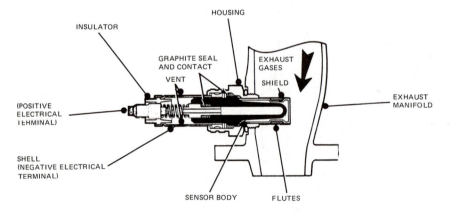

Figure 10-9 Ford O_2 sensor. (Courtesy of Parts and Service Division, Ford Motor Company.)

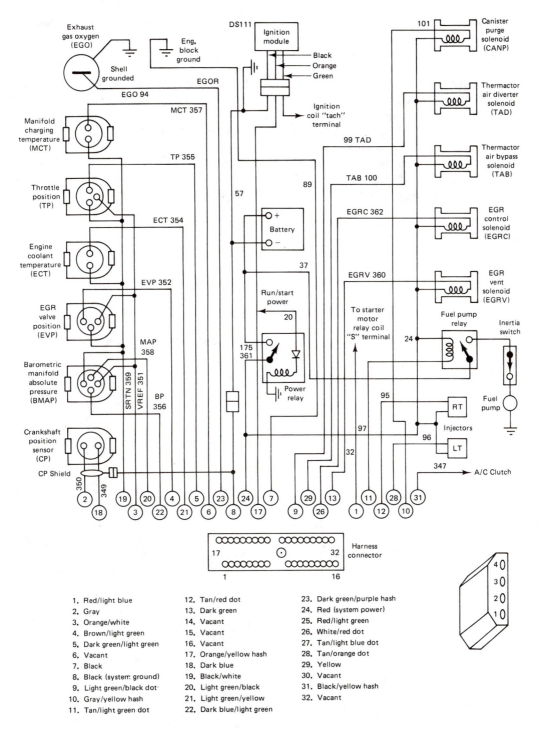

Figure 10-10 Ford EEC III wiring diagram. (Courtesy of Parts and Service Division, Ford Motor Company.)

1. Red/light blue
2. Gray
3. Orange/white
4. Brown/light green
5. Dark green/light green
6. Vacant
7. Black
8. Black (system ground)
9. Light green/black dot
10. Gray/yellow hash
11. Tan/light green dot
12. Tan/red dot
13. Dark green
14. Vacant
15. Vacant
16. Vacant
17. Orange/yellow hash
18. Dark blue
19. Black/white
20. Light green/black
21. Light green/yellow
22. Dark blue/light green
23. Dark green/purple hash
24. Red (system power)
25. Red/light green
26. White/red dot
27. Tan/light blue dot
28. Tan/orange dot
29. Yellow
30. Vacant
31. Black/yellow hash
32. Vacant

ELECTRICAL CONNECTORS

Snap-apart connectors are used to join the cables leading back to the control computer. In many cases, the male side of the connector has a release tab, which, when gently lifted, allows the two sides of the connector to be pulled apart. The connectors are joined by aligning the two halves, then pressing them together. The holding tabs provide the right tension for securing a good electrical connection.

Most connector terminals are not numbered. However, they are designed or shaped to connect only in the proper way. Sometimes they are color coded so that after testing one half, the unit can be put back together properly (Fig. 10-10).

11

Output Devices

Just as your hands perform actions in response to signals from your brain, output devices perform actions in response to signals from an on-board computer. The output devices physically manage the operation of the engine, directly controlling the air/fuel mixture, ignition timing, and certain emission controls. As time goes by, more and more engine and related operations will be handled by output devices controlled by impulses from a computerized brain. This chapter reviews the basic operating principles employed by most output devices, then examines some representative examples.

CONVERTING ELECTRICAL SIGNALS INTO PHYSICAL ACTIONS

Output devices must convert electrical signals into physical actions. Two of the most commonly used conversion mechanisms are solenoids and electric motors.

Solenoids

A typical solenoid consists of a coil (or coils) of wire wrapped around a hollow tube. When electrical current flows through the coil, a strong magnetic field is generated. The field, as it flows around and through the hollow tube, causes the solenoid to act like a magnet, with the ability to attract iron-based objects. However, unlike permanent magnets, a solenoid can be turned on or off very rapidly. A solenoid is often used to draw an object inside its hollow core. Such objects, referred to as *core rods*,

are attached by a linkage system to the mechanical device operated by the solenoid
(Fig. 11-1).

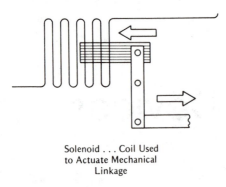

Solenoid . . . Coil Used
to Actuate Mechanical
Linkage

Figure 11-1 Solenoid. (From T.
Weathers and C. Hunter, *Diesel
Engines for Automobiles and Small
Trucks,* Reston Publishing Company,
Inc., Reston, Va., 1981.)

Electric Motors

Electric motors contain two sets of magnets, the field and the rotor (or armature).
The field may be an electromagnet or a permanent magnet. The rotor is always an
electromagnet.

Figure 11-2 illustrates a very simple electric motor. The field is a horseshoe
magnet and the rotor is a single loop of wire placed between the ends of the horseshoe magnet. The rotor is connected by a switch (called a split-ring commutator) to
the power source.

In terms of the imaginary magnetic particles described in Chapter 5, this is how
the motor works: Particles travel from the north to the south pole of the field magnet.
At the same time, particles circle the rotor loop, going around one way as the current

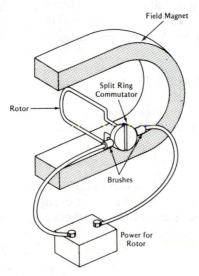

Figure 11-2 Electric motor. (From T.
Weathers and C. Hunter, *Diesel
Engines for Automobiles and Small
Trucks,* Reston Publishing Company,
Inc., Reston, Va., 1981.)

flows down one side the loop, and revolving the other way as the current flows back.

The end view of the motor shown in Fig. 11-3 shows how the particles interact to turn the rotor. The top part of the rotor is pulled to the left because the particles from the field and the rotor move in the same direction on the left side of the loop and in the opposite direction on the right side of the loop. The particles on the left try to join up, which pulls the top of the loop to the left. At the same time the particles on the right push apart, which pushes the top to the right.

A similar action takes place at the bottom of the loop. However, since the current (and hence the particles) move in the opposite direction, the pushing/pulling is to the right.

These forces move the rotor halfway around, as shown in Fig. 11-4. Momentum then takes the rotor on a little farther. However, unless something happens, the forces acting on the rotor will cancel out at this point. They will work in opposite directions and the rotor will stop.

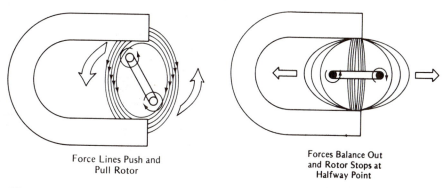

Force Lines Push and
Pull Rotor

Forces Balance Out
and Rotor Stops at
Halfway Point

Figure 11-3 End view of a motor showing the reaction between field and rotor force lines. (From T. Weathers and C. Hunter, *Diesel Engines for Automobiles and Small Trucks,* Reston Publishing Company, Inc., Reston, Va., 1981.)

Figure 11-4 Rotor after turning halfway around. (From T. Weathers and C. Hunter, *Diesel Engines for Automobiles and Small Trucks,* Reston Publishing Company, Inc., Reston, Va., 1981.)

That is where the split-ring commutator comes into play. As shown in Fig. 11-5, it reverses the current flow through the loop. The two halves of the commutator swap connections with the power source. What was once the negative connection to the rotor becomes the positive lead, and vice versa. As a result, the particles or lines of force circle the loop in opposite directions. Therefore, the forces acting on the rotor are reversed, causing it to continue on around. This cycle is repeated with every revolution of the rotor.

The remainder of this chapter elaborates on the mechanical operation of representative output devices from different manufacturers.

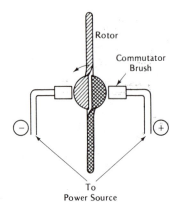

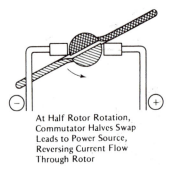

At Half Rotor Rotation,
Commutator Halves Swap
Leads to Power Source,
Reversing Current Flow
Through Rotor

Figure 11-5 Commutator reversing current flow. (From T. Weathers and C. Hunter, *Diesel Engines for Automobiles and Small Trucks,* Reston Publishing Company, Inc., Reston, Va., 1981.)

ELECTRONIC FUEL-INJECTION SYSTEMS

GM

The main components in the GM DEFI fuel-injection system (Fig. 11-6) include:

1. Fuel tank
2. Fuel pump
3. Fuel filter
4. Fuel pressure regulator
5. Fuel injectors

Fuel tank.   A special feature of the GM fuel tank is the reservoir located beneath the sending unit/in-tank pump assembly. Containing the fuel inlet and receiving returned fuel from the regulator, the reservoir is designed to provide a reliable source of fuel as long as any remains in the tank. Its shape and location also minimize the effects of the vehicle's maneuvering on fuel supply and delivery.

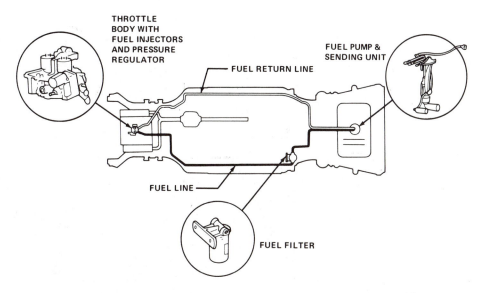

Figure 11-6 Fuel system components. (Courtesy of General Motors Corporation.)

Fuel pump. The in-tank pump is constructed integrally with the sending unit (Fig. 11-7). Using twin turbines driven by an electric motor, the pump supplies fuel under constant pressure to the filter and regulator.

Fuel filter. The fuel filter is mounted on the frame near the left rear wheel of most models. The unit is made of cast metal and has a replacement paper filtering element.

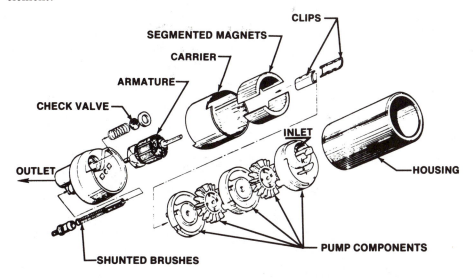

Figure 11-7 Fuel pump. (Courtesy of General Motors Corporation.)

Fuel pressure regulator.   The regulator is located in the throttle body (Fig. 11-8). It contains a spring-loaded, diaphragm-operated pressure relief valve. One side of the diaphragm is exposed to the atmosphere, the other side to fuel under pressure. The diaphragm and spring work together to maintain fuel pressure to the injectors at 10 psi over atmospheric pressure. Excess fuel is returned to the fuel tank.

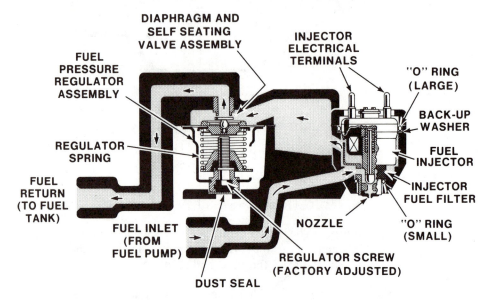

Figure 11-8   Fuel pressure regulator and injector. (Courtesy of General Motors Corporation.)

Fuel injectors.   The heart of the fuel injection system comprises the two fuel injectors located in the throttle body over the throttle plates. The main elements in each injector includes an injector body, solenoid coil, spring-loaded ball valve, valve seat/atomizer, injector fuel filter, and fuel inlet port (Fig. 11-9). This is how an injector works:

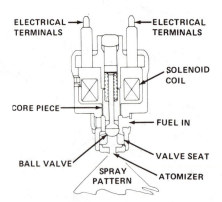

Figure 11-9   Fuel injector. (Courtesy of General Motors Corporation.)

Fuel is constantly delivered through the inlet port injector filter to a reservoir at the lower end of the injector body. When energized by a signal from the on-board computer, the solenoid lifts the spring-loaded ball valve off its seat. This opens up the injector, causing the pressurized fuel to spray through the atomizer into the airstream over the throttle plates. When the computer cuts off the signal to the solenoid, the spring-loaded ball valve closes, allowing a fresh charge of fuel to build up in the reservoir.

Ford Electronic Ignition System

Comparable elements of the Ford Electronic Fuel-Injection (EFI) system are shown in Figs. 11-10 through 11-13.

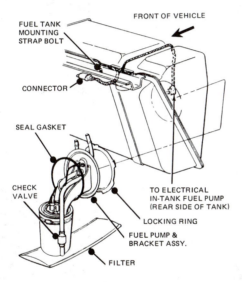

FRONT OF VEHICLE

FUEL TANK
MOUNTING
STRAP BOLT

CONNECTOR

SEAL GASKET

CHECK
VALVE

TO ELECTRICAL
IN-TANK FUEL PUMP
(REAR SIDE OF TANK)

LOCKING RING

FUEL PUMP &
BRACKET ASSY.

FILTER

Figure 11-10 Electric fuel pump for Ford EFI system. (Courtesy of Parts and Service Division, Ford Motor Company.)

CARBURETOR MIXTURE CONTROL DEVICES

At the time of this writing, many manufacturers use an on-board computer to control the output of a more-or-less conventional carburetor (although the trend is probably toward fuel-injector systems). The GM version, known as the Feedback Carburetor, employs a mixture control solenoid which operates a valve to determine the amount of fuel flowing through the main and idle circuits (Fig. 11-14 and 11-15). The computer controls the ground circuit to the solenoid. When the ground is complete, the energized solenoid closes off the valve to reduce the amount of fuel delivered and thereby lean out the mixture. The solenoid opens and closes the valve 10 times per second.

The Ford system, called the Feedback Carburetor Actuator, uses a solenoid

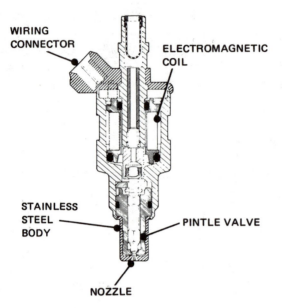

WIRING CONNECTOR

ELECTROMAGNETIC COIL

STAINLESS STEEL BODY

PINTLE VALVE

NOZZLE

Figure 11-11 Ford EFI fuel injector. (Courtesy of Parts and Service Division, Ford Motor Company.)

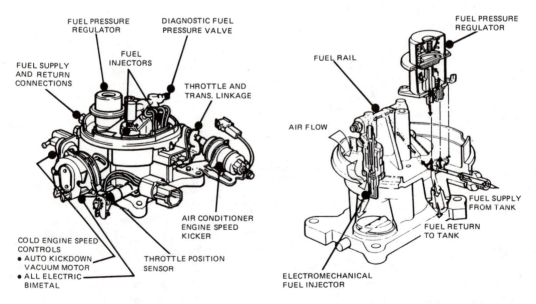

FUEL PRESSURE REGULATOR

DIAGNOSTIC FUEL PRESSURE VALVE

FUEL INJECTORS

FUEL SUPPLY AND RETURN CONNECTIONS

THROTTLE AND TRANS. LINKAGE

AIR CONDITIONER ENGINE SPEED KICKER

COLD ENGINE SPEED CONTROLS
• AUTO KICKDOWN VACUUM MOTOR
• ALL ELECTRIC BIMETAL

THROTTLE POSITION SENSOR

FUEL PRESSURE REGULATOR

FUEL RAIL

AIR FLOW

FUEL SUPPLY FROM TANK

FUEL RETURN TO TANK

ELECTROMECHANICAL FUEL INJECTOR

Figure 11-12 Ford EFI fuel charging system. (Courtesy of Parts and Service Division, Ford Motor Company.)

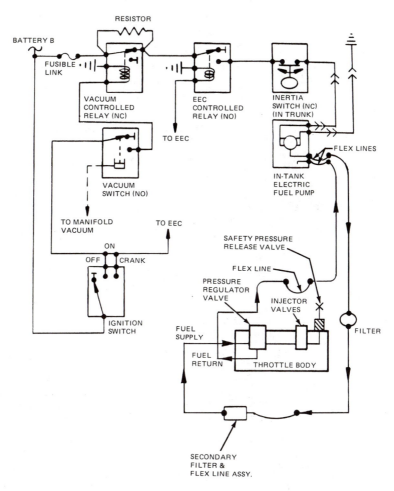

Figure 11-13 Ford EFI fuel pump wiring and fuel line diagram. (Courtesy of Parts and Service Division, Ford Motor Company).

stepper motor (Fig. 11-16). The motor has four separate armature windings or coils. Energized sequentially by the computer, these coils can move a control rod in and out to 120 positions or "steps." The total travel is 0.400 inch.

The stepper control rod is attached to a vacuum metering rod. The position of the metering rod determines the vacuum level applied to the float bowl, which, in turn, determines the pressure differential. The richest mixture occurs when the metering rod is fully extended, the leanest mixture when the rod is fully retracted.

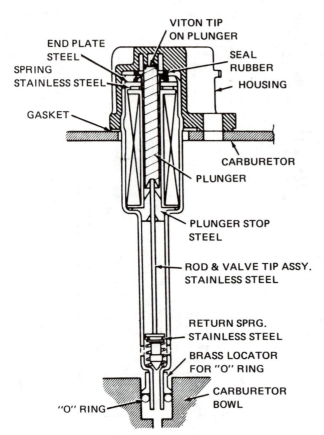

VITON TIP
ON PLUNGER

END PLATE
STEEL

SEAL
RUBBER

SPRING
STAINLESS STEEL

HOUSING

GASKET

CARBURETOR

PLUNGER

PLUNGER STOP
STEEL

ROD & VALVE TIP ASSY.
STAINLESS STEEL

RETURN SPRG.
STAINLESS STEEL

BRASS LOCATOR
FOR "O" RING

CARBURETOR
BOWL

"O" RING

Figure 11-14 GM mixture control solenoid (E2SE). (Courtesy of Chevrolet Motor Division, General Motors Corporation.)

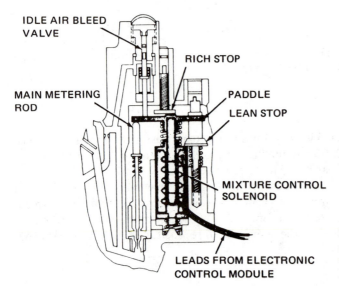

IDLE AIR BLEED
VALVE

RICH STOP

MAIN METERING
ROD

PADDLE

LEAN STOP

MIXTURE CONTROL
SOLENOID

LEADS FROM ELECTRONIC
CONTROL MODULE

Figure 11-15 GM mixture control solenoid (E2ME, E4ME). (Courtesy of Chevrolet Motor Division, General Motors Corporation.)

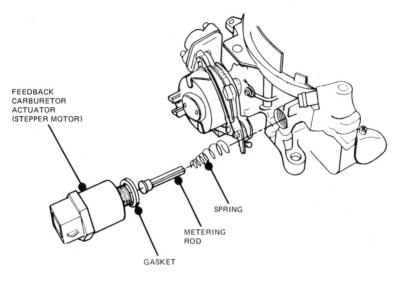

FEEDBACK
CARBURETOR
ACTUATOR
(STEPPER MOTOR)

SPRING

METERING
ROD

GASKET

Figure 11-16   Ford feedback carburetor actuator. (Courtesy of Parts and Service Division, Ford Motor Company.)

HIGH-ENERGY IGNITION SYSTEMS

The primary output components in most computer-controlled ignition systems are the distributor and ignition coil. Operating according to the control function identified in Chapter 9, the computer controls timing by interrupting the primary current flow to the ignition coil. The resulting high-voltage surges from the secondary windings are sent by the distributor to the spark plugs. In most cases, the distributor no longer contains springs or weights for advancing the spark (Fig. 11-17). That is handled by the computer. The dwell period is also managed by the computer.

Mechanically, then, computer-controlled distributors are likely to appear quite simple (Fig. 11-18). The most complicated aspects are the special circuitry associated with the distributor. For instance, the GM system employs a seven-terminal HEI (High-Energy Ignition) module in addition to the basic distributor (Fig. 11-19). The HEI module serves several functions:

1. It converts signals from the magnetic timer on the end of the crankshaft (described in Chapter 9) into reference signals which it then sends to the computer.
2. Upon receiving a firing signal from the computer, it interrupts the primary circuit in the ignition coil.
3. Under certain operating conditions (if the computer fails, if the engine is cranking, if the engine speed is below 200 rpm, or if the supply voltage to the computer drops below 9 V) the ignition module controls the ignition coil by itself.

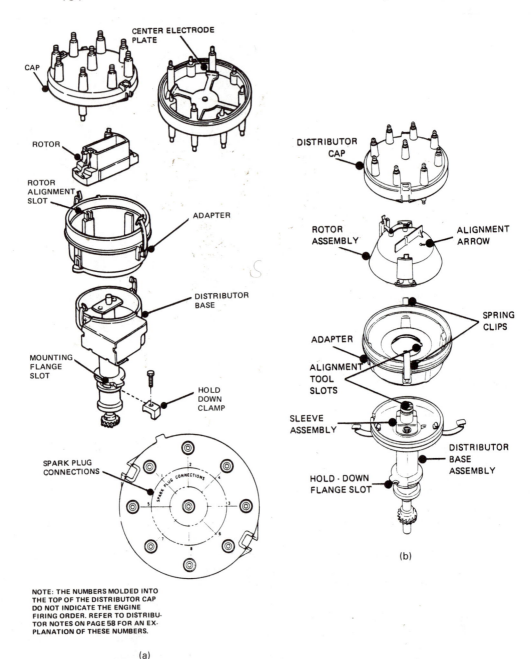

CENTER ELECTRODE PLATE

CAP

ROTOR

ROTOR ALIGNMENT SLOT

ADAPTER

DISTRIBUTOR BASE

MOUNTING FLANGE SLOT

HOLD DOWN CLAMP

SPARK PLUG CONNECTIONS

SPARK PLUG CONNECTIONS

NOTE: THE NUMBERS MOLDED INTO THE TOP OF THE DISTRIBUTOR CAP DO NOT INDICATE THE ENGINE FIRING ORDER. REFER TO DISTRIBUTOR NOTES ON PAGE 58 FOR AN EXPLANATION OF THESE NUMBERS.

(a)

DISTRIBUTOR CAP

ROTOR ASSEMBLY

ALIGNMENT ARROW

SPRING CLIPS

ADAPTER

ALIGNMENT TOOL SLOTS

SLEEVE ASSEMBLY

DISTRIBUTOR BASE ASSEMBLY

HOLD - DOWN FLANGE SLOT

(b)

Figure 11-17   (a) First-generation Ford electronic distributor; (b) second-generation Ford electronic distributor. (Courtesy of Parts and Service Division, Ford Motor Company.)

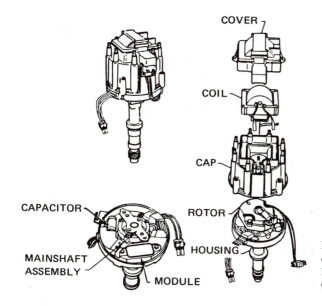

Figure 11-18 GM electronic distributor. (Courtesy of Chevrolet Motor Division, General Motors Corporation.)

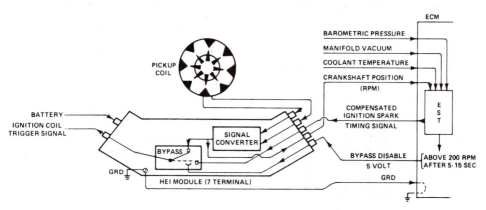

Figure 11-19 Schematic of GM electronic ignition system. (Courtesy of Chevrolet Motor Division, General Motors Corporation.)

AIR MANAGEMENT SYSTEM

As noted in Chapter 9, this system is used to provide extra air to the exhaust manifold or to the catalytic reactor (Fig. 11-20). Added O_2 (delivered on command from the onboard computer) helps oxidize HC and CO.

The main mechanical components of the system include a vane-type air pump (on many models) and a series of valves (Figs. 11-21 through 11-24). The air pump, which is driven by a belt and pulley connected to the end of the crankshaft, operates

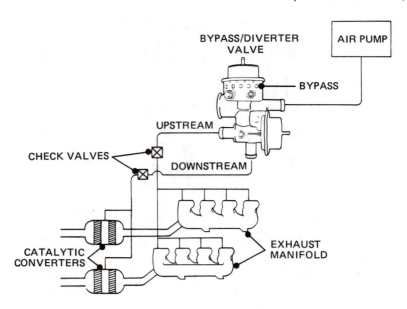

Figure 11-20 Thermactor air control valve for Ford EEC-III system. (Courtesy of Parts and Service Division, Ford Motor Company.)

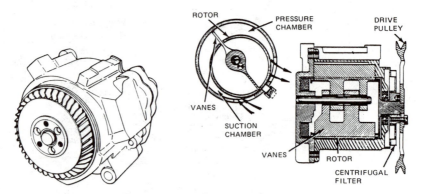

Figure 11-21 Thermactor air pump. (Courtesy of Parts and Service Division, Ford Motor Company.)

Figure 11-22 Thermactor air pump pressure/vacuum chambers. (Courtesy of Parts and Service Division, Ford Motor Company.)

all the time. The main valves are opened and closed by computer-controlled solenoids. The position of the main valves determines which components receive air. Additional check valves protect the system during deceleration and at other times when extreme pressures might cause damage.

In most versions of the system, a hierarchy of valves is used (Fig. 11-25). At the top of the hierarchy is a primary valve which directs the airflow to a secondary valve, or vents the air to the atmosphere (or to the air cleaner on GM models). The secondary valve sends the air to the exhaust manifold or to the catalytic reactor.

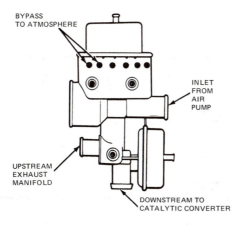

Figure 11-23 Thermactor bypass/diverter valve. (Courtesy of Parts and Service Division, Ford Motor Company.)

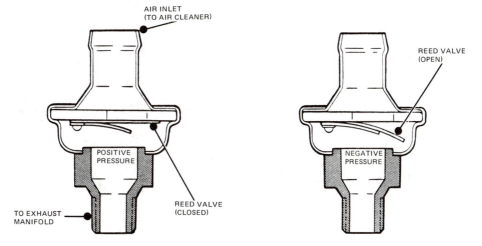

Figure 11-24 Thermactor air check valves. (Courtesy of Parts and Service Division, Ford Motor Company.)

IDLE SPEED CONTROL

The idle speed output control determines the position of the throttle plate(s). In the 1981 GM system, a small electric motor is mounted on the side of the throttle body (Fig. 11-26). The motor operates a gear and shaft assembly, which in turn, determines the closing stop position of the throttle plates.

The Ford system works in a different manner. Instead of using a motor, it employs a solenoid-operated throttle kicker. Responding to signals from the on-board computer, the kicker controls the position of the vacuum actuator diaphragm. When the solenoid is engaged, the diaphragm extends the throttle stop to increase engine idle speed.

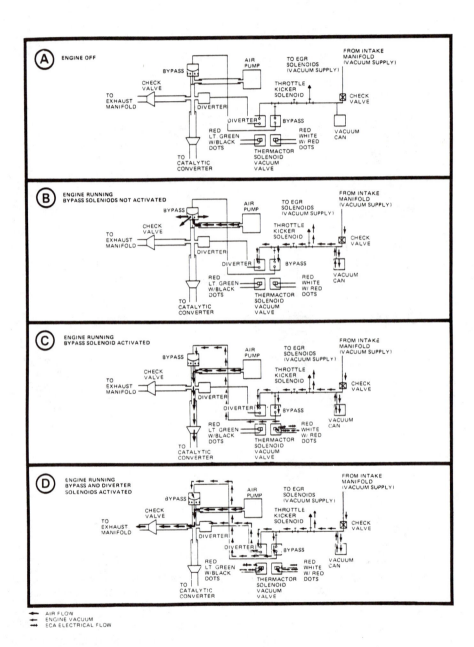

Figure 11-25   Thermactor air system schematic. (Courtesy of Parts and Service Division, Ford Motor Company.)

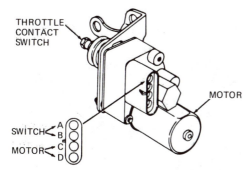

Figure 11-26 GM idle speed control motor. (Courtesy of Chevrolet Motor Division, General Motors Corporation.)

GM MODULATED DISPLACEMENT ENGINE

The GM modulated displacement system features solenoid-operated output devices to control the operation of four sets of intake and exhaust valves. The cylinders affected are numbers 1, 7, 4, and 6 (Fig. 11-27). During certain conditions (described in Chapter 9), valve opening for cylinders 1 and 4 is stopped, leaving six cylinders operating. When other conditions are sensed by the on-board computer, the valves on all four cylinders are stopped, resulting in four-cylinder operation.

Valve opening is controlled by adjusting the position of the rocker pivot point or fulcrum (Fig. 11-28). During normal operation, when the solenoid is not energized, a blocker plate pushes down on a pivot spring located over the end of the rocker arm. When the pushrod forces one end of the rocker arm up, the middle of the arm pivots about the pivot spring and pushes down on the valve stem and valve spring.

However, when the solenoid is energized, the blocker plate is rotated. Holes in the blocker plate line up with projections on top of the pivot spring. No longer

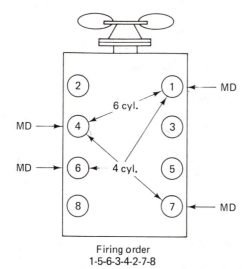

Firing order
1-5-6-3-4-2-7-8

Figure 11-27 Modulated displacement cylinder diagram. (Courtesy of General Motors Corporation.)

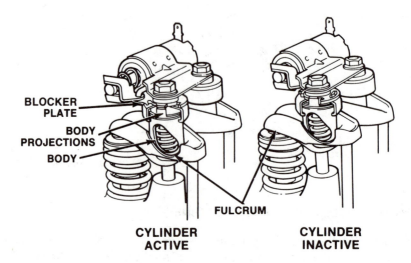

Figure 11-28   Valve selector hardware. (Courtesy of General Motors Corporation.)

held down by the blocker plate, the pivot spring expands. It still exerts enough pressure to keep the pushrod and valve train components from flopping about. However, it does not offer enough resistance to form a fulcrum for the rocker arm. The fulcrum point shifts to the outside end of the rocker arm; consequently, it is no longer able to open the valve.

In addition to the changes directly associated with the output units, modulated displacement engines are different from conventional engines in other ways (Fig. 11-29). The camshaft lobes have been modified to compensate for the extra lash (play)

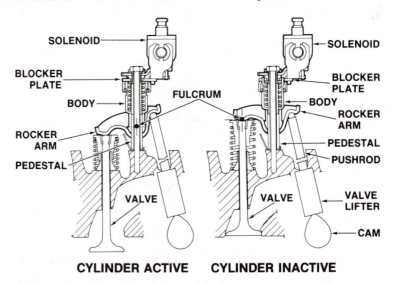

Figure 11-29   Selector operation. (Courtesy of General Motors Corporation.)

in the selector assembly. Rocker arm covers have been redesigned to give enough room for the valve actuators. Because there are not as many cylinders operating, a belt-driven pump is needed to power vacuum-operated accessories.

TORQUE CONVERTER CLUTCH

This GM system (called TCC) controls a solenoid-operated clutch mounted in the automatic transmission (Figs. 11-30 and 11-31). Under certain conditions (described in Chapter 9), the onboard computer energized the solenoid, causing the clutch to couple the engine directly to the transmission. When the operating conditions indicate that the engine should operate in a normal fluid coupling manner, the solenoid and clutch are disengaged.

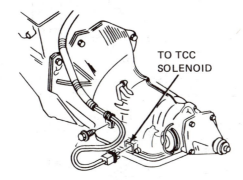

Figure 11-30 Transmission clutch control solenoid. (Courtesy of Chevrolet Motor Division, General Motors Corporation.)

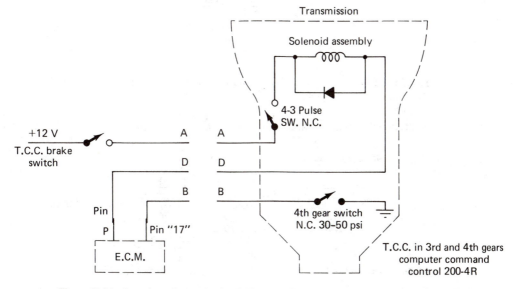

Figure 11-31 Location of control solenoid in transmission. (Courtesy of Chevrolet Motor Division, General Motors Corporation.)

EXHAUST GAS RECIRCULATION (EGR) CONTROL VALVES

EGR valves control the ported vacuum to the EGR valve (Fig. 11-32). The EGR valve itself is usually not directly manipulated. It responds to the vacuum source the same as it did before computerized controls. The control valves are opened and closed by computer-controlled solenoids (Fig. 11-33). When the solenoids are energized, the valves close, thereby shutting off the vacuum source to the EGR valve. The EGR valve cannot operate at that time. However, when the solenoids are deenergized, the control valves open. Manifold vacuum then determines the operation of the EGR valve (Fig. 11-34 through 11-36).

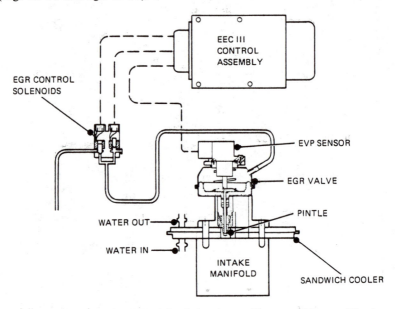

Figure 11-32 Ford exhaust gas recirculation system. (Courtesy of Parts and Service Division, Ford Motor Company.)

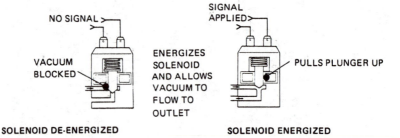

Figure 11-33 Ford EGR control valve solenoid. (Courtesy of Parts and Service Division, Ford Motor Company.)

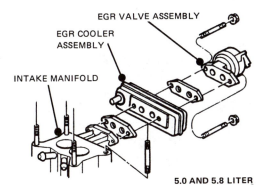

EGR VALVE ASSEMBLY

EGR COOLER
ASSEMBLY

INTAKE MANIFOLD

5.0 AND 5.8 LITER

Figure 11-34 Ford EGR gas cooler. (Courtesy of Parts and Service Division, Ford Motor Company.)

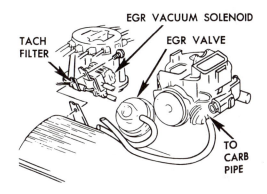

EGR VACUUM SOLENOID

TACH
FILTER

EGR VALVE

TO
CARB
PIPE

Figure 11-35 GM EGR solenoid. (Courtesy of Chevrolet Motor Division, General Motors Corporation.)

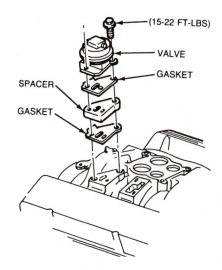

(15-22 FT-LBS)

VALVE

GASKET

SPACER

GASKET

Figure 11-36 GM EGR mounting assembly. (Courtesy of Chevrolet Motor Division, General Motors Corporation.)

EARLY FUEL EVAPORATION (EFE) CONTROL VALVE

The EFE system is used to supply extra heat to the incoming air and fuel when the engine is cold. The extra heat helps fuel evaporate, thereby promoting more complete combustion. This, in turn, improves performance and reduces emissions.

At the time of this writing, GM is one of the few manufacturers using an on-board computer to control EFE. Two systems are employed. One system uses a solenoid-controlled vacuum actuator motor and an exhaust heat valve. The valve is located between the exhaust manifold and the exhaust pipe (Fig. 11-37). The computer-con-

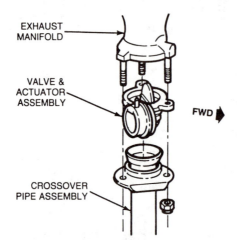

EXHAUST MANIFOLD

VALVE & ACTUATOR ASSEMBLY

FWD

CROSSOVER PIPE ASSEMBLY

Figure 11-37 Early fuel evaporation valve. (Courtesy of Chevrolet Motor Division, General Motors Corporation.)

trolled solenoid determines the position of the valve by regulating the supply of vacuum to the actuator motor.

The other EFE system, used on some smaller, carburetor-based engines, employs a ceramic heater grid located under the primary bore of the carburetor (Fig. 11-38).

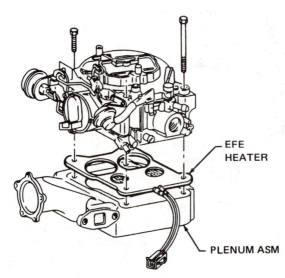

EFE HEATER

PLENUM ASM

Figure 11-38 Early fuel evaporation heater. (Courtesy of Chevrolet Motor Division, General Motors Corporation.)

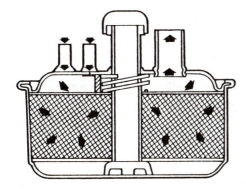

Figure 11-39 Charcoal canister.
(Courtesy of Parts and Service Division,
Ford Motor Company.)

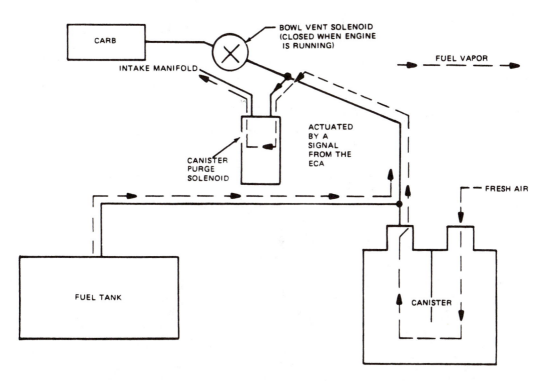

Figure 11-40 Canister purge schematic. (Courtesy of Parts and Service Division, Ford Motor Company.)

When the ignition switch is turned on and the coolant temperature is low, voltage is applied by the computer to a relay connected to the heater grid. The relay closes the circuit to the heater, which then heats the incoming air and fuel. After the temperature increases to a certain point, the computer deenergizes the relay, which shuts off the heater circuit.

CANISTER PURGE VALVE

A solenoid-operated valve is used in many systems to control the flow of excess fuel vapors from a charcoal storage canister (Fig. 11-39). When the solenoid is energized by the on-board computer, the control valve opens. Trapped gases can then be forced from the canister to the lower-pressure region inside the intake manifold (Fig. 11-40).

12

Review of Meters and Circuit Testing

Meaningful testing has always depended on using the correct tools in the proper manner. This is especially true when checking computer-related systems. A hit-or-miss casual approach simply will not work. More problems will be created than solved.

The last two chapters in the book, this one and the next, deal with testing. In the present chapter we review some common types of test devices and general testing procedures. The final chapter will examine specific procedures for testing computer-related systems.

THREE BASIC TYPES OF ELECTRICAL TEST METERS

As you are probably already aware, the three basic types of electrical test meters are voltmeters, ammeters, and ohmmeters. In many instances, these meters are combined into one case with switches for selecting the desired function. These units are called *multimeters*. A number of meters use a needle pointer or "hand" to portray readings. Other meters use digital readout screens. The next several paragraphs review the operation of dial-type meters.

GENERAL CONSTRUCTION AND OPERATION

All dial-type meters are constructed in a similar manner. The basic movement includes a permanent fixed-position magnet and a movable electromagnet with a needle attached. Current from the test circuit flows through the electromagnet, causing it to be surrounded by lines of magnetic force. These force lines interact with the force lines surrounding the permanent magnet. The resulting attracting and repulsion causes the electromagnet and attached needle to move in proportion to the current flowing through the electromagnet.

To ensure accuracy, the electromagnet is balanced between two jeweled bearings. The electromagnet is also spring loaded so that the attached needle rests (or is "pegged") toward one end of the scale (normally the left).

Current for the electromagnet comes from two flexible leads which are attached to the circuit being tested. Current variations in the test circuit alter the ampere-turns of the electromagnet, thus affecting its strength. The strength of the electromagnet determines the way it responds in the presence of the permanent magnet, which, in turn, determines the position of the needle.

The outside of the meter contains various scales and control knobs. If the meter is a multifunction unit, controls will be provided for selecting either voltmeter, ammeter, or ohmmeter functions. Most meters also provide controls for selecting the test range best suited for the conditions being measured. Turning the selector knob changes the resistance of the input circuit of the electromagnet, thereby altering the response of the magnet to the circuit being tested.

Test leads are usually identified as being positive or negative. Positive leads are colored red and/or marked with a plus (+) sign. Negative leads are colored black and/or marked with a negative (−) sign. Various kinds of ends are found on test leads. Some have sharp, probe tips; others have alligator clips. The particular application for a meter (voltmeter, ammeter, ohmmeter) is determined both by the internal wiring of the meter and by the way it is attached to the circuit being tested.

Voltmeters

A typical voltmeter is pictured in Fig. 12-1. Voltmeters are always attached in parallel to the circuit being tested. In other words, the test circuit is left completely hooked up and the voltmeter leads are connected on either side of the element being tested.

Ammeters

Figure 12-2 shows a typical ammeter. Notice the two shunt branches used to determine the test range, one branch for testing 0 to 6 amperes and the other circuit for checking 0 to 30 amperes. Aside from these internal differences, one of the most significant distinctions between an ammeter and voltmeter is the way in which the test leads are attached to a circuit. Ammeter leads are connected in series with the test circuit so that all the current being measured flows through the meter.

Ohmmeters

Unlike voltmeters and ammeters, ohmmeters are *never* hooked into a "live" circuit. To do so could cause damage both to the meter and the circuit. Instead, an ohm-

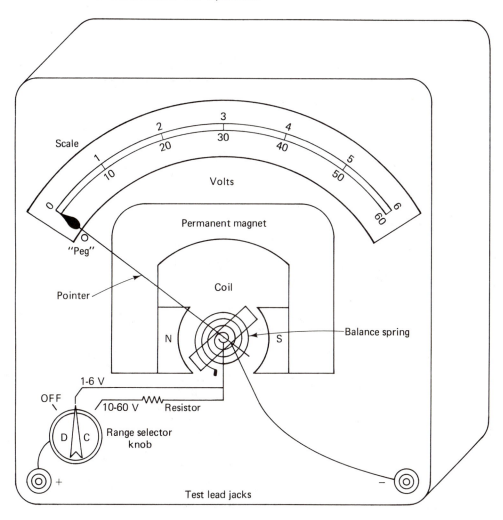

Figure 12-1 Voltmeter.

meter, as pictured in Fig. 12-3, has its own power source, often a regular penlight battery. The battery sends a known current flow through the test leads into the detached ends of the circuit being tested. The resistance of the circuit affects the flow of current back into the meter, which, in turn, determines the behavior of the electromagnet and attached needle. Each time an ohmmeter is used, it must be calibrated. Ohmmeters made by different manufacturers are sometimes used or calibrated in different ways, so be sure to check the operating instructions before employing an unfamiliar unit.

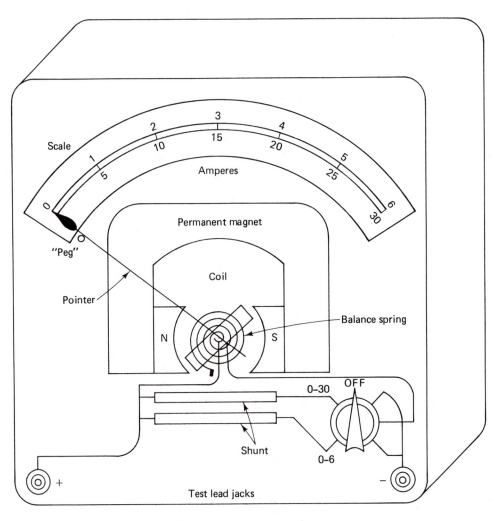

Figure 12-2   Ammeter.

DIGITAL METERS

The most obvious difference between digital and dial-type meters involves the way in which test results are displayed. Dial-type meters are analog devices, representing test results on a divided scale. Digital meters, on the other hand, display specific readings, often accurate to +0.1%. Internally, digital meters are also different from dial-type meters. Instead of using two sets of magnetic fields, digital meters use analog-to-digital converters, similar in some cases to the ones used in on-board computers. Because of these differences, digital meters are often more accurate than dial-type meters and can be used over a wider range. For this reason, critical, computer-related

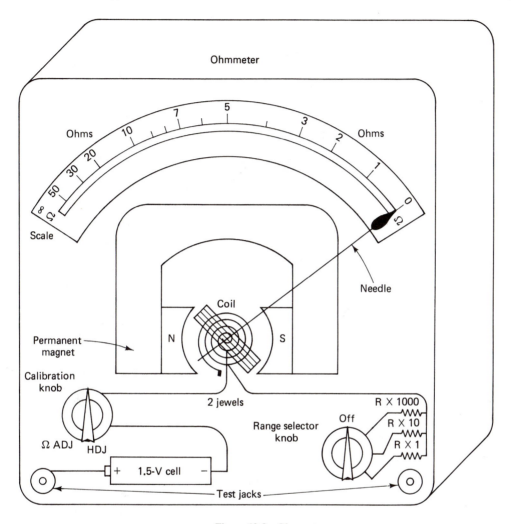

Figure 12-3 Ohmmeter.

testing sometimes requires a digital volt/ohmmeter with a minimum independence range of 10 megaohms (Fig. 12-4).

OTHER TYPES OF TEST DEVICES

Test Lights

Test lights, which have been used for years, are among the most common electrical test gear. They are still used for some computer-related diagnosis.

A test light is basically two wires connected to a light bulb. Some mechanics

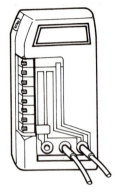

Figure 12-4   Digital volt/ohm meter.

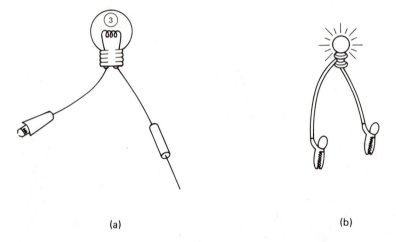

(a)   (b)

Figure 12-5   Unpowered test lights.

make their own. Unpowered test lights are used to check for power (Fig. 12-5). When connected in parallel or series to a live circuit, the light will usually come on. Powered test lights, which contain a small pen-light battery, are used to check for continuity (Fig. 12-6). The leads are connected to both ends of the circuit or element being checked. If the circuit is complete, the bulb will light up.

Test Leads and Jumper Wires

These components (Fig. 12-7 and 12-8) are used to extend the range of a meter's lead or selectively to bypass (or short out) a section of circuit. They are often used in diagnosing computer-controlled devices when circuits must be disconnected and meters or test lights must be connected at various locations.

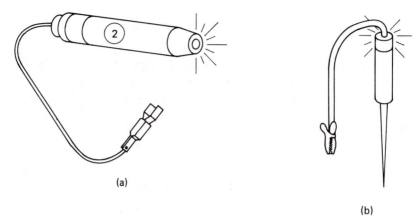

(a)

(b)

Figure 12-6 Powered test lights.

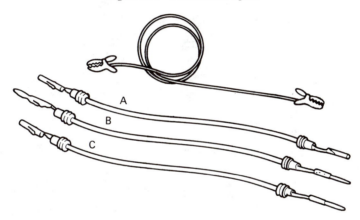

Figure 12-7 Clip jumper wire and test leads.

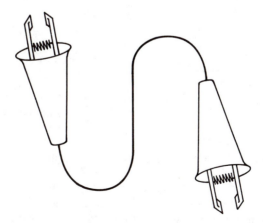

Figure 12-8 Jumper cable.

Dwell Tachometer

This familiar test device (Fig. 12-9) is also used in computer-related testing. The tachometer, as in the past, gives engine speed in rpm. However, as we will see in Chapter 13, the dwell scale is no longer used to check ignition dwell. It is employed for other purposes. That is because dwell is not adjustable in computer-controlled ignition systems.

Vacuum Pump

Hand-operated vacuum pumps (Fig. 12-10) are often employed in testing air- or vacuum-operated devices, such as the EGR and air management systems.

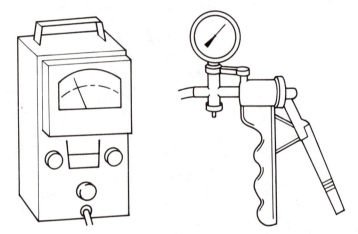

Figure 12-9 Tach/dwell meter.

Figure 12-10 Vacuum pump.

HOW TEST DEVICES ARE USED

For the remainder of this chapter, we will look at some common applications of the test devices just introduced. Chapter 13 describes actual computer system testing and diagnosis.

Testing Headlights with Unpowered Test Light

We begin with the simplest type of test device, the unpowered test light. In this exercise, the tester is being used to check for current flow up to the headlights (Fig. 12-11). If the test light fails to come on, you would know something is wrong on the positive side of the headlight circuit.

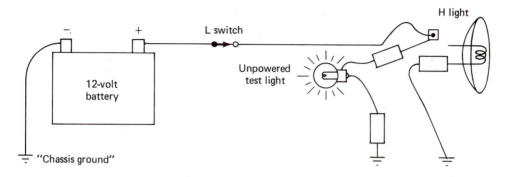

Figure 12-11   Testing for power at the headlight using an unpowered test light.

Testing Headlights with a Powered Test Light

The powered test light in this example is used to determine if the negative side of the headlight circuit is complete (Fig. 12-12). If there are no breaks or regions of excessively high resistance, current will flow and the test light will shine. Such a test can be used to check continuity in almost any circuit.

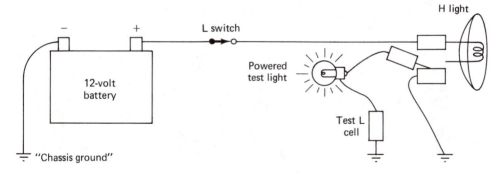

Figure 12-12   Testing the ground circuit of the headlight using a powered test light.

Testing the Ignition Coil with an Ohmmeter

The ohmmeter in this test is being used to check the resistance of a coil's primary windings (Fig. 12-13). As noted earlier, the ohmmeter has its own power source and is never connected to a live circuit. Here is the general procedure for using common types of ohmmeters:

1. Calibrate the meter by placing the prods together and turning the adjusting knob until the needle is at the proper calibration point (usually zero).
2. After calibrating the meter, place the tips of the prods against the terminals of the circuit being tested.

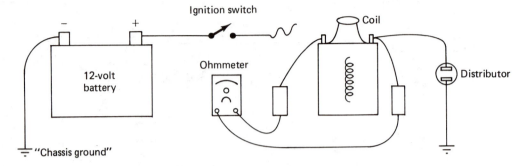

Figure 12-13 Measuring the resistance of the primary windings of the ignition coil with the switch off and the coil disconnected.

3. Unless you know the general resistance to expect, turn the selector knob to the lowest range, usually R × 1 (the resistance reading times 1, in other words, the actual reading on the scale).

4. If the meter hand moves off the scale, adjust the selector until the needle stays within the scale. If that turns out to be R × 10, you would multiply the scale reading times 10. If the adjustment is R × 100, you would multiply the scale reading times 100.

Testing the Ignition Coil with a Voltmeter

A voltmeter lets you determine the electrical pressure anywhere along a circuit. In this example, the voltmeter is being used to check the electrical pressure in an old, points-style ignition system. Although the test is no longer valid, it does illustrate the principle of voltmeter testing. Notice in Fig. 12-14 that the distributor side of the coil has been grounded. This is to prevent the power in the circuit from going on and off as the engine is turned over by the starter motor (and the points open and close).

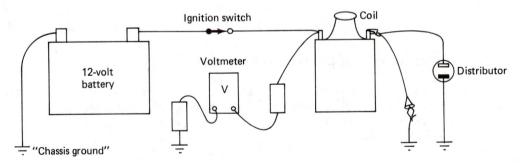

Figure 12-14 Measuring the voltage at the ignition coil with the switch on, the cranking motor turning over, and the distributor side of the coil grounded.

Checking the Heater Motor with an Ammeter

In this test, an ammeter is used to check the current flow through a heater motor circuit (Fig. 12-15). As noted before, an ammeter must be hooked in series with the circuit being tested so that all the current flows through the meter. In the example shown here, the leads from the switch to the motor have been disconnected so that the ammeter can be connected in series with the motor. Another point to remember is not to exceed the ammeter rating. Do not try to test circuits producing more current than the ammeter can handle.

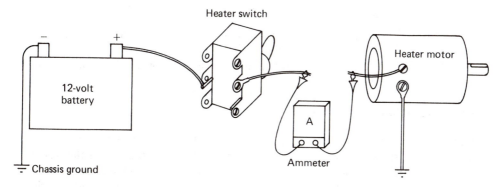

Figure 12-15   Checking current flow with an ammeter connected in series.

Checking the Starter Motor Circuit with a Voltmeter

Here we have a voltmeter connected on either side of a starter motor solenoid. It is being used to check the resistance in the circuit. You might think that an ohmmeter should be used. However, since the starter motor is the largest user of battery energy, the cable has a very low resistance, on the order of 0.0008 ohm in some cases. Many ohmmeters do not have scales that read this low. The solution is to check for voltage, since it will increase when ohms diminishes and decrease when ohms goes up.

This test reading is known as the *voltage drop*. It is an indication of the pressure needed to overcome the resistance of a given section of circuit. Voltage drop specifications are provided by many manufacturers for various circuits. A general rule of thumb says that each cable or switch in many circuits will have a voltage drop of 0.1 volt. All the voltage drops must add up to the total voltage impressed on the circuit.

The test shown in Fig. 12-16 is conducted in this manner:

1. Select the lowest scale on the voltmeter.
2. Connect the meter leads as shown in the figure.
3. Temporarily disconnect the ignition so that the car will not start.

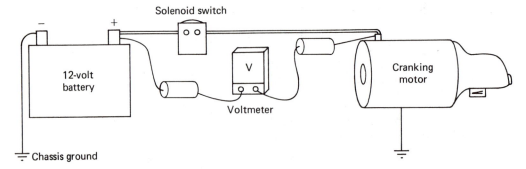

Figure 12-16   Using a voltmeter to check voltage drop with the engine cranking over.

4. With the engine cranking over, read the voltmeter scale. The reading will be the voltage drop.

Checking Continuity with an Ohmmeter

Continuity can be checked in many ways. Here an ohmmeter is used to check the continuity of the cables in a wiring harness (Fig. 12-17). To perform the test, first calibrate the meter as noted before. Then, after placing the range knob in the R × 1 position, place one of the tester probes on one of the cable connectors. To determine if the cable is complete, place the other probe at the other end of the cable. A complete cable will allow current to flow and will produce a reading on the tester scale. This reading can be compared to the manufacturer's specifications.

At this point, you might be wondering why an ohmmeter is required for continuity testing—why, for instance, a powered test light is not used. Simply this: If the resistance in the circuit is high, enough current might not pass for the bulb to shine, even though the resistance is within specifications. This is especially true for test lights using a 1½-volt power cell.

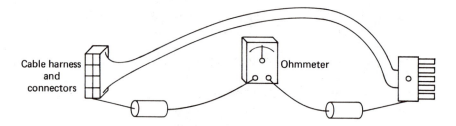

Figure 12-17   Using an ohmmeter to check for continuity and conductor resistance.

Checking Continuity with a Nonpowered Test Light

In this test, a nonpowered test light is used to check for continuity through the terminals in a switch (Fig. 12-18). While one probe remains in contact with ground, the other probe is moved from terminal to terminal. The test light will not shine when

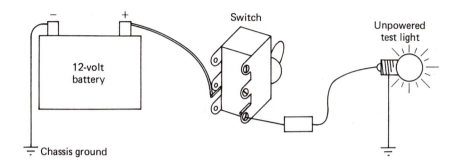

Figure 12-18 Using an unpowered test light to check for continuity through a switch.

connected to an incomplete circuit (assuming that the switch is on). The same general procedure can be used to check continuity in any circuit. Starting with the connection nearest the battery, move the tester probe from connection to connection away from the battery. The first connection where the bulb fails to shine indicates a malfunction.

Checking Vacuum Leaks with a Hand-Operated Vacuum Pump

Not all computer-related testing is done with electrical test gear. It is sometimes necessary to check the operation of vacuum- or air-operated output devices. In this example, the pump is used to "pull" a vacuum on the air switching unit (Fig. 12-19).

Figure 12-19 Hand-operated vacuum pump used to check vacuum operated actuators, vacuum lines, and switches.

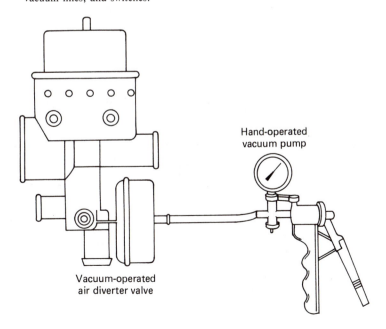

Then the leakage rate is determined by observing the vacuum gauge to see how long it takes for the vacuum to be dissipated. If the rate exceeds the manufacturer's specifications, a problem is indicated.

Vacuum Gauges

A vacuum gauge such as the one shown in Fig. 12-20 can be connected anywhere to check vacuum leaks.

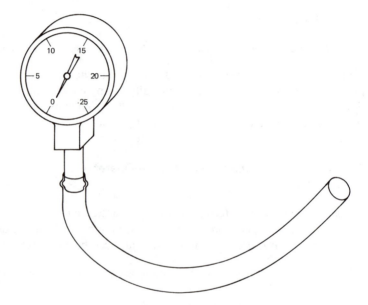

Figure 12-20   Vacuum gauge.

13

Diagnosis and Testing

This chapter should be especially useful for working mechanics or for anyone who expects to diagnose and test computer-controlled engine systems. Although complete diagnostic procedures for specific engines are not covered, general directions and guidelines are provided. These tips will help you develop the correct approach for dealing with computerized systems. They will also help you understand and use diagnostic materials supplied by the automotive manufacturers.

FIND THE CORRECT APPROACH

There have always been two basic approaches to automotive problem solving and repair. We can call one the "shade tree" approach and the other the "logical" approach. Before getting into the discussion of diagnosis and testing, it will be helpful to think about these two approaches and how they relate to you.

Shade Tree Approach

Shade tree work is based on the mechanic's experience and on a certain "feel" for automobiles that comes from that experience. Test equipment and detail specifications are used only when absolutely necessary (usually, not even then). Operational theory, especially the kind that comes out of books, is generally ignored. The shade tree mechanic prefers to discuss engine operation (if he or she talks about it at all) in nontechnical, everyday terms. Some shade tree mechanics may be found under literal shade trees (Fig. 13-1). However, many others work in garages, service stations, and dealerships.

In the past, being a shade tree mechanic was not necessarily such a bad thing. This was especially true when it came to relying on the tricks and skills picked up from years of experience. A good shade tree mechanic could often get a car back in operation long before a "book-learned" novice could even diagnose the problem.

Unfortunately, the shade tree approach is not adequate for modern computer-controlled systems. The equipment is so complex and it changes so often that it is no longer possible to develop a "feel" based strictly on working experience. Without knowing how the systems operate, it is difficult to even get any working experience.

Figure 13-1   Shade tree operation.

Logical Approach Required

The second, logical approach is needed today (Fig. 13-2). This does not mean that the good shade tree operator did not do logical work. What it does mean is that the instinctive, subconsciously applied logic of the shade tree will not work in situations where the mechanic has no direct working experience. Nowadays, the logic must be applied beforehand and must be based on an understanding of the system(s) involved and a thorough study of the problem.

However, Mechanic's Logic Not Enough

At this point, the observant reader may have spotted what appears to be a flaw in the foregoing argument. This reader might say: "If a logical, problem-solving approach requires an understanding of the subject, how can I ever hope to fix computer-controlled systems? You've already told me that I can't really expect to understand

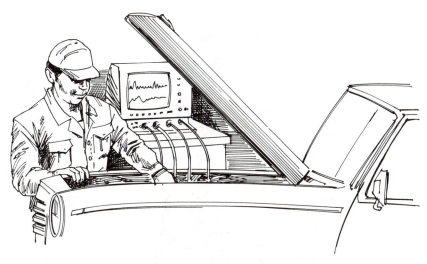

Figure 13-2 Logical approach.

computers without some special study. Does this mean I can't diagnose problems in computer-controlled systems?''

Strangely enough, the answer is "yes." Unless you happen to be an expert in automotive computer-controlled engine systems, you cannot diagnose problems on your own. You need help from someone who is an expert. Help comes in the form of diagnostic charts, tables, and guides contained in the manufacturers' manuals (Fig. 13-3). Prepared by system experts, these materials are an absolute ''must'' for effective diagnosis and repair of computer-controlled systems. They supply the expertise that you cannot be reasonably expected to possess by yourself.

Of course, the observant reader might now make another comment. "OK, if the shop manuals are going to tell me how to diagnose and test problems anyway, why do I need this book? Why do I need to know anything about computer-controlled systems?''

Quite frankly, some experts would answer that mechanics do not need to know much, certainly not as much as this book covers. They would say that it is asking too much of you and is a waste of time. They would also add that you are better off simply following the directions on a diagnostic chart, even if you do not know what you are doing or why.

Obviously, the authors of this book do not agree. First, no diagnostic chart, no matter how complete, can cover every situation and problem. There will be times when you need to work at least partially on your own. To have any hope of success in these situations, you will need some background information. Also, the manufacturers' diagnostic manuals are not always easy to follow. This is understandable, since they deal with complex subjects. So just to use these manuals, you must possess some knowledge of the subject being discussed.

Figure 13-3   Diagnostic guides and manuals. (Courtesy of Chrysler Corporation, Ford Motor Company and General Motors Corporation.)

INFORMATION REQUIRED

Before doing any sort of diagnosis you need information about the problem. Usually, this information comes from a number of sources:

1. Customer comments
2. Visual inspection
3. On-board diagnostics
4. Testing

Customer Comments

The first place to go for information is person who has firsthand knowledge of the problem: the customer. You must actually listen to what the customer says (Fig. 13-4). The report may be garbled and confusing, and perhaps contradictory, but it often holds the key to the problem. Ideally, after listening to the customer's initial comments, you will think for a moment, come up with a preliminary diagnosis, then ask some pertinent questions to help confirm or reject your ideas. Of course, listening to the customer is always a good practice, whether the problem relates to a computer-controlled system or to some other part of the vehicle.

Figure 13-4 Listening to customer comments. (Courtesy of General Motors Corporation.)

Visual Inspection

The next information-gathering activity is the visual inspection. It only makes sense to eliminate obvious sources of difficulty before getting into more complex testing and diagnosis. As before, this recommendation applies to any sort of problem, whether or not it relates to a computerized system.

The following is a checklist of activities that might be performed during the course of a typical visual inspection.

1. Remove the air cleaner and inspect for dirt or foreign material or other contamination in and around filter element.
2. Examine vacuum hoses for proper routing and connection. Also check for cracked broken or pinched hoses or fittings.
3. Examine each portion of the computer wiring harness. Check for the following at each location.
   (a) Proper connections at sensors and solenoids
   (b) Loose or disconnected connectors
   (c) Broken or disconnected wires
   (d) Partially seated connecters
   (e) Broken or frayed wires
   (f) Shorting between wires
   (g) Corrosion
4. Inspect each sensor for obvious physical damage.
5. Operate the engine and inspect the exhaust manifold and exhaust gas oxygen sensor for leaks.
6. Repair faults as necessary, and reinstall the air cleaner.

On-Board Diagnostics

As noted in Chapter 9, some computerized systems have built-in, self-diagnostic programming. These programs check to make sure that input is received from all the sensors and that it lies within a predefined range. The operation of certain output actuators may also be checked. The particular self-diagnostic features vary from manufacturer to manufacturer.

The results obtained by the on-board diagnostics are compared with values stored in the computer's permanent ROM or PROM memory. If any results are out of specification, the computer makes note of the fact.

At the time of this writing, GM and Ford take two different approaches to on-board diagnostics. GM systems perform ongoing testing whenever the engine is running. If a problem is encountered, "CHECK ENGINE LIGHT" flashes on the dash, letting the driver know that something is wrong. The appropriate trouble code is also stored in the terminal's RAM memory.

The trouble code remains in RAM as long as the power remains on to the computer. That is because the RAM area is like a slateboard or scratch pad. Whenever current stops flowing through the tiny circuits and switches, all information stored in RAM disappears.

To see what codes (if any) are stored in memory, the mechanic grounds a test lead under the dash. Trouble codes will then appear as patterns of flashes of the test light. Once all the trouble codes have been displayed, the mechanic looks them up in the diagnostic guides provided by GM.

Ford, on the other hand, does not conduct ongoing diagnostic checks. Diagnostic programs (which are stored in the PROM calibration ships) are initiated only after the mechanic performs certain actions to the engine. Once started, the diagnostic programs take over the operation of the engine, directing it through preplanned test routines. If problems are encountered, service codes are signaled by patterns of thermistor solenoid pulses. When the special EEC tester is used, the pulses will flash on the tester's display panel. If the tester is not used, the operation of the solenoids themselves must be observed. After the test is complete, the Ford mechanic looks up the service codes in the appropriate diagnostic reference guides. (Details of both GM and Ford self-diagnostic features are covered later in this chapter in the section "Special Manufacturer Notes.")

As you might imagine, these trouble codes can be very valuable aids in locating problems in computerized systems. However, it should be understood that not all possible defects are covered. Also, it must be understood that trouble codes cannot pinpoint problems with absolute certainty. Trouble codes are simply one category of information to be examined during the course of carrying out the manufacturer's diagnostic procedure.

Test Results

This is the final information category we discuss before getting into diagnostic procedures. Test results are obtained by using test instruments in a manner prescribed by the manufacturers. Usually, the test procedure is included in an overall diagnostic sequence. The results of a test are compared to the manufacturer's specifications. The outcome is then used as basis for deciding what path to follow in the remaining diagnostic procedure, or the result could form the basis for making repairs.

The particular test equipment used depends on the manufacturer and on the test being performed. Common test devices used by GM and Ford are pictured in Figs. 13-5 and 13-6.

DIAGNOSTIC AND TEST PROCEDURES

First, let us see what is meant by a diagnostic procedure. If you scan the manufacturer's guides, charts, checklists, and so on, you will notice that they contain tests and questions regarding various aspects of the vehicle's operation. Depending on the answers to these questions and the results of the tests, you will move in one direction or another through the diagnostic procedure. The ultimate objective is to pinpoint a particular

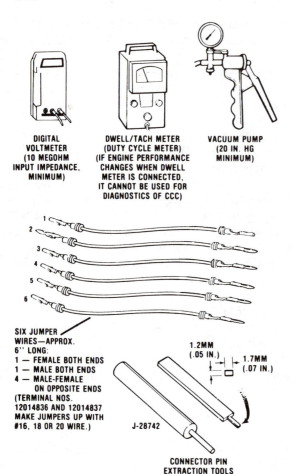

DIGITAL
VOLTMETER
(10 MEGOHM
INPUT IMPEDANCE,
MINIMUM)

DWELL/TACH METER
(DUTY CYCLE METER)
(IF ENGINE PERFORMANCE
CHANGES WHEN DWELL
METER IS CONNECTED,
IT CANNOT BE USED FOR
DIAGNOSTICS OF CCC)

VACUUM PUMP
(20 IN. HG
MINIMUM)

SIX JUMPER
WIRES—APPROX.
6'' LONG:
1 — FEMALE BOTH ENDS
1 — MALE BOTH ENDS
4 — MALE-FEMALE
    ON OPPOSITE ENDS
(TERMINAL NOS.
12014836 AND 12014837
MAKE JUMPERS UP WITH
#16, 18 OR 20 WIRE.)

1.2MM
(.05 IN.)

1.7MM
(.07 IN.)

J-28742

CONNECTOR PIN
EXTRACTION TOOLS

Figure 13-5 GM test equipment.
(Courtesy of Delco Electronics Division,
General Motors Corporation.)

component that requires repair or replacement. In some cases, the defective component is identified early in the procedure. In other cases, the procedure must be followed all the way to the end. In certain instances, the procedure does not apply at all and the problem will not be pinpointed.

Most diagnostic procedures have two principal features: they are hierarchical and branching (Fig. 13-7). By *hierarchical* we mean that the procedure starts out by examining general conditions, then, by the process of elimination, narrows the possibilities down to one culprit. *Branching* refers to the way you move through the diagnostic charts, tables, or steps. Rather than proceed in a straight line, you are likely to jump back and forth through the material as possible causes of the problem are eliminated.

Because of the hierarchical and branching features, no two diagnostic procedures are likely to follow exactly the same steps. However, up to a point, the overall diagnostic sequence is similar for most computer-controlled systems.

General Procedures

1. Usually, you start out by listening to the customer complaint. A tentative diagnosis of the problem may be made at this time.
2. A visual inspection of the engine compartment is performed next. Particular attention is paid to the components associated with the tentative diagnosis made earlier.

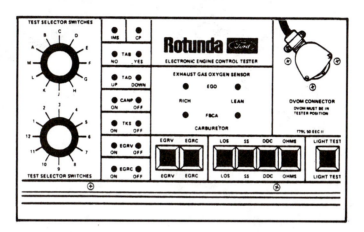

(a)

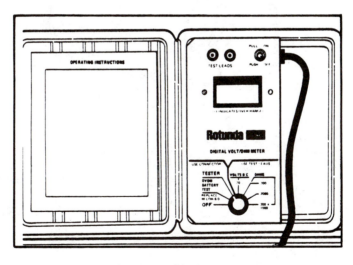

(b)

Figure 13-6 (a) Rotunda diagnostic tester; (b) digital volt/ohm meter; (c) tachometer; (d) vacuum gauge; (e) vacuum hand pump/tester; (f) special fuel injection tester harness; (g) fuel injection tester harness. (Courtesy of Parts and Service Division, Ford Motor Company.)

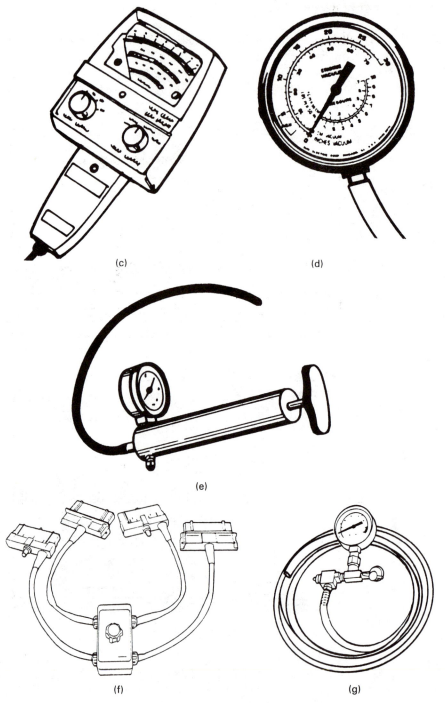

(c)

(d)

(e)

(f)

(g)

Figure 13-6 (*continued*)

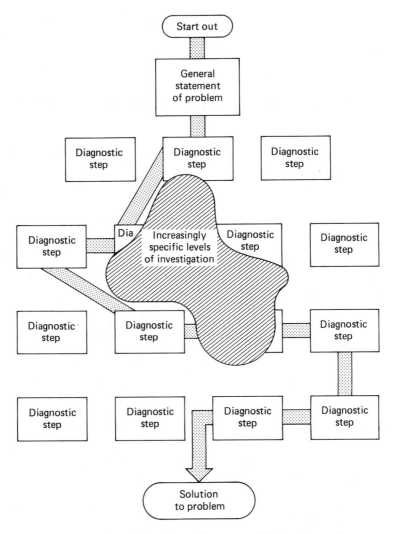

Figure 13-7 Hierarchical and branching nature of diagnostic procedures.

3. Any defects spotted in the visual inspections are repaired.
4. Since many problems are not caused by malfunctions in computer-related systems, all other possibilities are examined and any defects repaired.

Manufacturers' Procedures

From now on, the diagnosis concentrates on computer-related systems. The particular procedures followed will depend on the manufacturer's recommendations.

GM procedure.   At the time of this writing, GM starts all computer-related diagnosis and testing with a Diagnostic Circuit Check (Fig. 13-8). This set of procedures checks to make sure that:

1. The engine self-diagnostic system is working.
2. If the self-diagnostic system is working, what trouble codes, if any, are present.
3. If no codes are present, what other general problems may be present.

Depending on the results of the circuit check, the technician may be directed to:

1. Detail diagnostic checks for trouble codes encountered
2. Repair procedures to replace a defective ECM
3. Service procedures to fix loose or defective connectors
4. Another generalized diagnostic sequence, called the System Performance Check

If directed to the System Performance Check, the mechanic will work through another series of diagnostic steps that examine additional aspects of the ECM and related components. During the course of this procedure, the mechanic might end

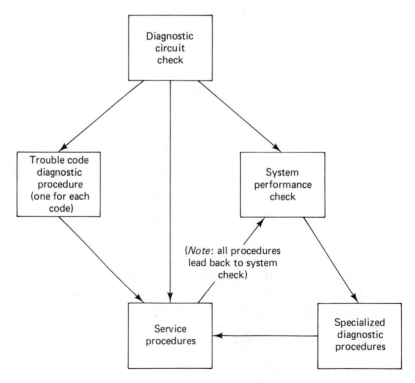

Figure 13-8 GM diagnostic sequence.

up performing repairs, replacing the ECM, or proceeding to additional, specialized diagnostic checks.

At the conclusion of any diagnostic procedures, when all the repairs have been made and the engine seems to be working properly, GM recommends that the System Performance Check be carried out one more time to make sure that the components are indeed working properly.

Ford EEC III diagnosis.   After a visual inspection, Ford testing generally proceeds to one of three diagnostic procedures: (1) a Self-Test for engines that will run, (2) a No-Start Test for carburetor-equipped engines what will not run, or (3) a Diagnostic Chart for fuel injector-equipped engines that will not run (Fig. 13-9).

1. *Self-Test*. This procedure may employ the Rotunda T791-50-EECII Diagnostic Tester. The tester is hooked up to the engine after the vehicle has been operated long enough for the radiator hose to become hot and pressurized. Then, with the engine running at idle speed, a hand-operated vacuum pump is attached to the barometric sensor vent outlet. The sensor vacuum is pumped down to 20 inches Hg

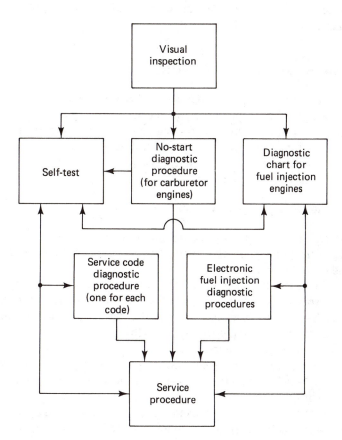

Figure 13-9   Ford diagnostic sequence.

and held there for 5 seconds. This reading, which is below any possible normal barometric pressure, causes the test cycle to begin (with or without the tester hooked up).

Responding to programs running in the vehicle's on-board microcomputer, the engine goes through a series of controlled operations. Information going to and from the sensors and output devices are compared with preset values stored in the computer. The results of this comparison, whether "OK" or out-of-spec, are signaled to the technician by thermactor solenoid pulses. Particular pulse patterns relate to certain service codes. After the test cycle has been completed, the technician checks the service codes against a diagnosis guide in the tester's operating manual.

2. *No-Start Test.* These diagnostic procedures check for computer-related conditions that might cause the no-start condition. The No-Start Test will lead the technician to the appropriate repair operation and (if the operation is successful) back to the self-test procedure for final confirmation of the work performed.

3. *Diagnostic Chart.* This procedure checks for a spark from the ignition coil and for fuel coming from the injectors. Depending on the results, the mechanic is directed to the No-Start Test or to an additional set of procedures called the Electronic Fuel Injection Diagnosis.

USING A DIAGNOSTIC REPAIR MANUAL

All of these procedures are described in detail in the manufacturer's diagnosis and testing manuals. Of course, as you may have already discovered, it is not always easy to use one of the manufacturer's manuals. You may understand individual sentences and paragraphs, but the whole thing still may not quite fit together. You may have trouble deciding where to get started or where to turn for the procedure that applies to your particular problem.

Multiple Levels of Information

One secret for understanding the manufacturer's materials is to realize that most diagnostic and repair manuals deal with several levels of information at the same time (Fig. 13-10).

Most manuals include at least some reference or background information. In other words, the manual explains the components covered in the diagnostic procedures. Sometimes the explanation will be "in-depth" and may include detail specifications for various types of engines. However, usually, that sort of detailed information is covered in manuals devoted strictly to repairs.

Reference or background information does not necessarily tell you how to do anything—simply how things work. It is the next level that tells you how to perform tasks—in this case, how to perform diagnostic procedures.

The "how-to" directions in diagnostic manuals may take several forms. Some how-to instructions are simple numbered steps. Other instructions are organized in-

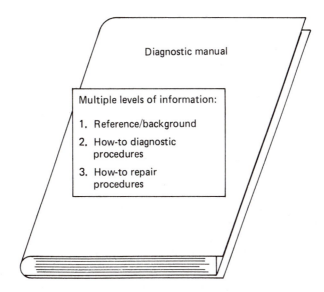

Figure 13-10 Diagnostic manuals contain multiple levels of information.

to table form, where each test or question, the possible results, and the mechanic's actions are described in separate columns. Other how-to instructions appear as charts, resembling program flowcharts.

Occasionally, diagnostic manuals may include a third level of information. This level constitutes how-to instructions for conducting actual repair procedures. However, most repairs are not explained in detail. It is assumed that the mechanic either knows how to perform the task, or that it is covered in some other document.

The important point to remember is that the different levels of information require different approaches by the reader. When you read reference or background information, you are trying to see how components work and how they relate to one another. You may also be looking for particular facts about a component whose basic operation you already understand. So, you probably will not read every word, only those passages that relate to your particular needs. The organization of reference material usually lends itself to this sort of information-gathering process.

How-to guides, on the other hand, are meant to be read word for word and followed exactly in the order presented. You cannot skip around through the material; you must do exactly what the directions tell you to do.

The trouble with many diagnostic and testing manuals is that they mix reference and how-to information without giving the reader a clear idea about the organization of the different parts. As a result, the wrong approach may be taken to the manual. In particular, how-to directions may not be followed exactly. So, given the fact that manuals may be confusing, it is up to you to figure out what is what and to decide on the correct approach to the different parts.

Representative Manuals

To help you in this task, we examine briefly parts of several representative manuals.

GM "Pocket Brains".   This small booklet is used as a quick guide to the 1981 GM Computer Command Control (CCC) system. It contains both reference and how-to information.

As indicated by the table of contents pictured in Fig. 13-11, the manual is divided into three major parts. The first part consists of pictures noting the locations and relationships among the major components (Fig. 13-12). Pictures of the major pieces of test equipment used are also provided. These pictures, which are given without much additional explanation, serve as ready-reference material for people who already have some idea of the operation of the system.

The second part of the booklet contains reference/background information with some how-to directions mixed in. Included are general descriptions of the diagnostic procedures, brief explanations of the CCC functions and components, and a description of the trouble code memory. How-to directions are also provided for checking out and replacing the plug-in PROM memory chip. Depending on your prior knowledge, you might skim over this part of the booklet, carefully read selected passages, or read the entire section in detail.

The third part of the booklet contains the actual diagnostic procedures. This section starts out with an introduction to the diagnostic process, then follows with a list of trouble codes and what they mean. The next few pages provide a mix of step-by-step and "tree" charts. The step-by-step directions describe the action to take in response to customer complaints. The first tree chart pictures the steps in the Diagnostic Circuit Check described earlier in this chapter (Fig. 13-13). The second tree chart starts out the System Performance Check.

From the first two tree charts, you are directed to the other charts contained in the remainder of the manual. Some of these charts tie back into the start of the procedure. Others stand alone or lead to other diagnostic procedures.

As you can tell, the organization and relationship of the material in the third part of the book is extremely complicated. This is where the shade tree and logical approaches part company for good. At this point, you must trust the developer of the diagnostic procedure. You must read the directions carefully and go from one test to the next exactly as you are told.

Note: To help use tree charts such as those pictured in the GM booklet, imagine that the chart is similar to a tree turned *upside* down on the page (Fig. 13-14). Everyone starts at the main trunk on the top of the page. The first test or question represents the first main fork in the tree. That fork is followed by another fork, and that fork by another. Depending on your responses or test results you will eventually end up on a branch that directs you to a certain procedure or to another diagnostic tree altogether.

Ford "Service Highlights" (EEC-III).   This manual is more reference oriented than the GM booklet described previously. As noted in the table of contents pictured

Contents

Figure 13-11 Table of contents from "Pocket Brains." (Courtesy of Delco Electronics Division, General Motors Corporation.)

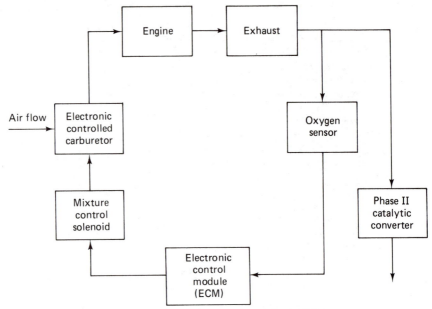

Block diagram of computer command control (CCC)
system for normal operation (fuel control)

Note: Oxygen sensor input is not used by the ECM
to make decisions during open loop operations.

(a)

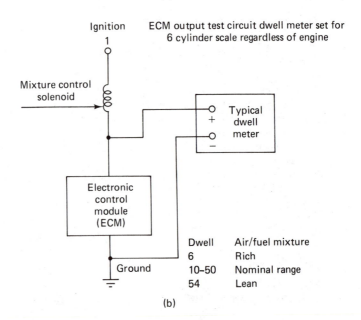

ECM output test circuit dwell meter set for
6 cylinder scale regardless of engine

Dwell	Air/fuel mixture
6	Rich
10–50	Nominal range
54	Lean

(b)

Figure 13-12 Pictures in the first part of "Pocket Brains." (Courtesy of Delco Electronics Division, General Motors Corporation.)

in Fig. 13-15, four of the six main sections deal with system description and operation features. Another section (the last) provides how-to instructions for performing certain repair operations. Only one section (the second) deals specifically with diagnostic information.

The diagnostic section itself is broken into six main parts. The first part pictures and briefly describes the test equipment required. The next part shows how to hook up the test equipment. The third part tells how to perform the visual inspection. The last three parts describe, in order: the Self-Test procedure, the No-Start Test, and the Electronic Fuel Injection Diagnosis. It is interesting to note that each diagnostic procedure is treated in a different way in the manual.

The Self-Test procedure is actually more of a reference guide. Mechanics are directed to the Test Operator's Manual for specific diagnostic procedures. However, mechanics who do not have access to that manual can extract enough information to do a certain level of diagnosis.

The No-Start procedure is organized into what is sometimes called a "step-go-to-step" table (Fig. 13-16). The table is divided into three columns, headed STEP, RESULT, and ACTION. The operations described in the STEP column are numbered and each describes a particular diagnostic test or question. The next column lists the possible RESULTS of each step. The third, ACTION, column directs the mechanic to another numbered STEP, to a particular repair operation, or to the Tester Operator's Manual. Most actions eventually branch back to step 1, to confirm that any repairs made have been successful.

The Electronic Fuel Injection Diagnosis contains commands and directions arranged in a paragraph and list format. Major groups of related activities are grouped under common headings. Most of the directions are branching; that is, you are directed to go from one step or heading to another depending on the results of your testing.

This last diagnostic guide may appear to be the easiest because it is not organized into an unfamiliar format. However, it is actually more complicated because it contains more descriptive text and because the connections between the various "branches" are not illustrated graphically. However, like any diagnostic guide, the best approach is to start at the beginning and to follow the directions wherever they lead you.

SPECIAL MANUFACTURER NOTES

The following paragraphs list special features of the GM and Ford systems mentioned previously in this chapter. For complete descriptions of these and other systems, refer to the manufacturers' manuals.

GM

1. *Trouble code display.* The dash-mounted "CHECK ENGINE LIGHT" flashes on whenever the computer's on-board diagnostics find a problem. If the problem is intermittent the light will go out, although the trouble code will remain in the

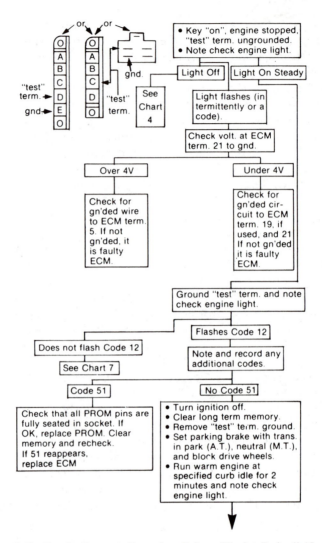

Figure 13-13 Sample diagnostic "tree charts" from "Pocket Brains." (Courtesy of Delco Electronics Division, General Motors Corporation.)

RAM memory area as long as power remains on to the computer.

To display trouble codes stored in memory, the mechanic grounds a test lead under the dash while the engine is running. The dash light will then flash on and off. By observing the sequence of the flashes, the mechanic can tell what trouble code is stored. For instance, if the light flashes two times, pauses for a moment, then flashes three more times, the mechanic knows that trouble code 3 has been stored (indicating a problem in the mixture control solenoid). Each pattern for a particular

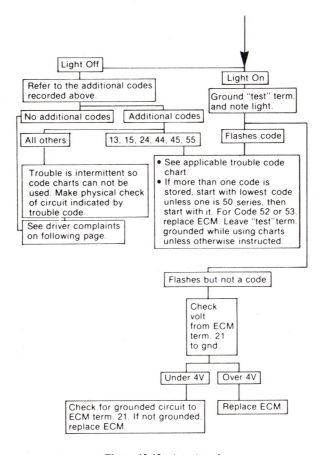

Figure 13-13 (*continued*)

trouble code flashes three times before the pattern for the next trouble code flashes (assuming that other trouble codes are stored).

2. *Checking the trouble code display.* There are several built-in checks of the trouble light bulb and circuit. The light should always come on when the ignition switch is turned to the "on" position with the engine not running. In addition, trouble code 12 (one flash, a pause, and two flashes) should be displayed whenever the test lead is grounded, the engine is not running, and the ignition switch is in the "on" position. (Trouble code 12 means something else when the engine is running.)

3. *Long-term versus short-term memory.* GM uses both of these terms discussing RAM memory storage. Here is what they mean: *Long-term memory* refers to the RAM storage provided by a computer whose "R" terminal has been connected directly to the battery. Supplying constant voltage to the computer ensures that information stored in RAM will remain available (as long as the power is available). *Short-term memory* refers to the RAM storage provided by a computer which is not connected

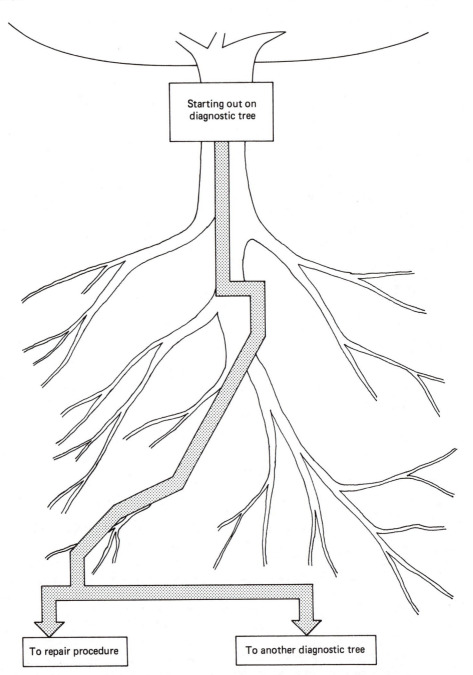

Figure 13-14 Using a diagnostic tree chart.

TABLE OF CONTENTS

Figure 13-15 Table of contents from Ford diagnostic manual. (Courtesy of Parts and Service Division, Ford Motor Company.)

Step	Result	Action
1 Attempt to start engine	Engine starts	Go to self test for system check
	Engine does not start	Go to 2
2 Check vehicle battery (VBAT) voltage · Set tester switches to "A" and "1" · Record battery voltage	10.5 volts or more	Voltage ok; go to 3
	Less than 10.5 volts	Voltage too low; stop testing — go to tester operator's manual

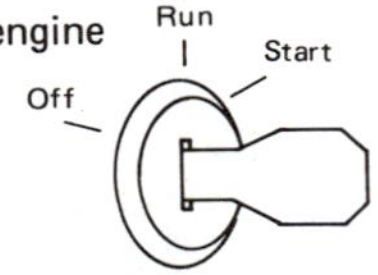

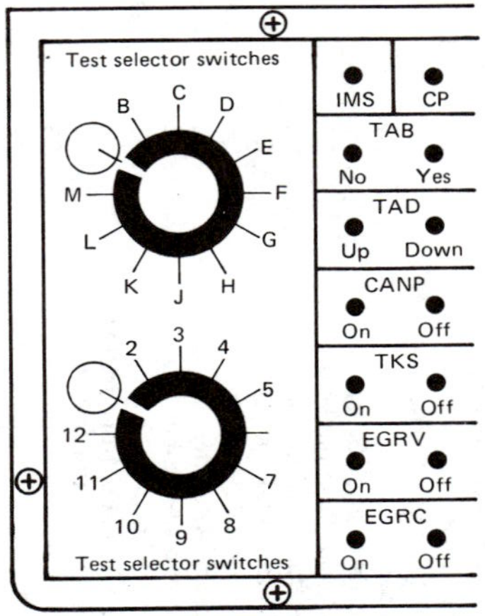

(a)

Figure 13-16   Sample "step-go-to-step" diagnostic guides. (Courtesy of Parts and Service Division, Ford Motor Company.)

Step	Result	Action
3 Check EGR valve position (EVP) sensor voltage: · Set tester switches to "A" and "g" · Record voltage	1.95 volts Not 1.95 volts	Voltage ok; go to ☐4 Voltage out of limits; go to tester operator's manual

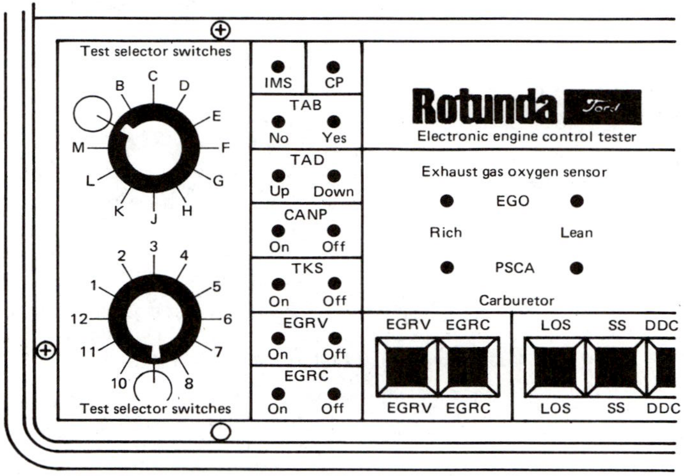

(b)

Figure 13-16 *(continued)*

235

Step	Result	Action
4 · Crank engine and observe ignition module signal ("IMS") and crankshaft position ("CP") lamps Note: An IMS or CP pulse (light on less than several seconds) when the key is first turned to start does not indicate a true computer output signal 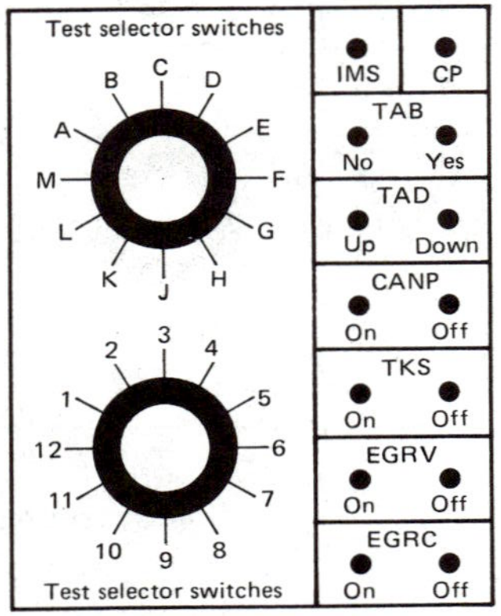Test selector switches A B C D E F G H J K L M 1 2 3 4 5 6 7 8 9 10 11 12 Test selector switches IMS CP TAB No Yes TAD Up Down CANP On Off TKS On Off EGRV On Off EGRC On Off 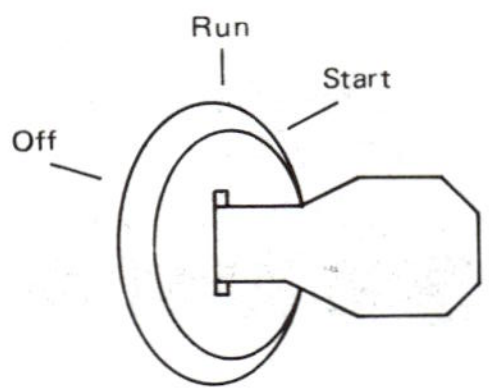Off Run Start	Both lights lit and spark	Go to tester operator's manual
	Both lights lit and no spark	ECA ok; go to [11] to verify circuit to ignition module
	"CP" light lit, "IMS" light not lit	ECA not providing signal to ignition module. Go to [11]
	Both lights not lit	No crankshaft position (CP). Sensor signal being received by ECA. Go to [5] to check CP sensor.
	"IMS" light lit, CP light not lit*	No crankshaft position (CP). Sensor signal being received by ECA. Go to [5] to check CP sensor.

Note: If engine will start now with "IMS" light lit and CP light not lit a CP circuit with a low output but one with enough power to start the car is indicated. This condition can cause a vehicle to fail to start in cold weather or make starting difficult. Proceed to [5] to check sensor circuits.

(c)

Figure 13-16 *(continued)*

permanently to the battery. Power in this case is available only when the ignition switch is on. Usually, the R terminal is not connected at the factory. This prevents battery drain in vehicles that might not be operated for long periods. The R terminal must be connected by the dealer. Also, after any testing the repair, the R terminal should be disconnected briefly. This will clear out trouble codes stored in RAM.

4. *Electrical connections.* Good electrical connections are important in any electrical system, especially an electronic system involved in complex information exchanges. Figure 13-17 pictures some of the GM computer system electrical harnesses and connectors and the location of the major components.

5. *PROM Installation.* The PROM memory unit, which contains calibration values for particular vehicles, is one of the few computer components that may be serviced at the local level. Usually, if the computer is replaced, the PROM chip is removed and reused on the replacement computer. Because it is so important to computer operation, the mechanic must be very careful to install the PROM chip correctly and securely. Figure 13-18 pictures some aspects of PROM installation.

6. *Delayed trouble codes.* Most trouble codes associated with the oxygen sensor will not show up immediately after the engine is cranked. It takes about 5 minutes before the sensor warms up. It also takes several minutes for codes associated with the coolant sensor to appear. A list of GM trouble codes are shown in Fig. 13-19.

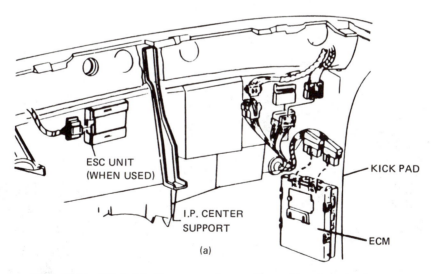

Figure 13-17 GM electrical harnesses and connections and locations of major components: (a) ECM mounting location, typical; (b) ECM terminal locations; (c) ALDL connector mounting, typical; (d) CCC component locations, typical; (e) oxygen sensor mounting typical; (f) pressure sensor mounting, typical; (g) ESC detonation sensor mounting, typical. (Courtesy of Delco Electronics Division, General Motors Corporation.)

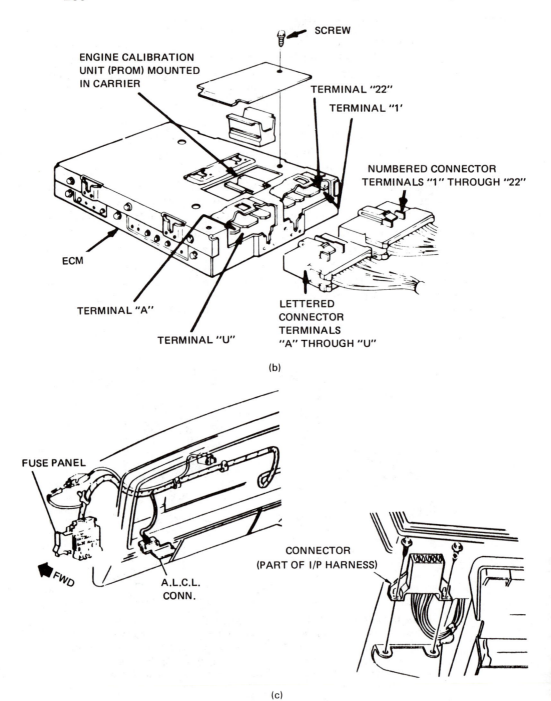

SCREW

ENGINE CALIBRATION
UNIT (PROM) MOUNTED
IN CARRIER

TERMINAL "22"

TERMINAL "1'

NUMBERED CONNECTOR
TERMINALS "1" THROUGH "22"

ECM

TERMINAL "A"

TERMINAL "U"

LETTERED
CONNECTOR
TERMINALS
"A" THROUGH "U"

(b)

FUSE PANEL

FWD

A.L.C.L.
CONN.

CONNECTOR
(PART OF I/P HARNESS)

(c)

Figure 13-17   (*continued*)

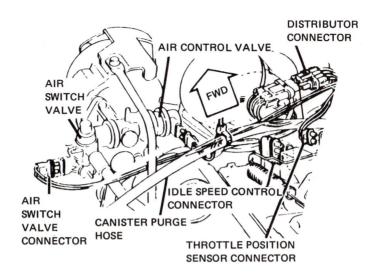

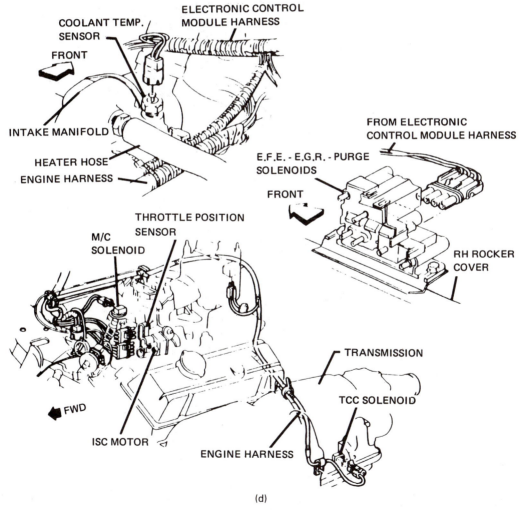

(d)

Figure 13-17 (*continued*)

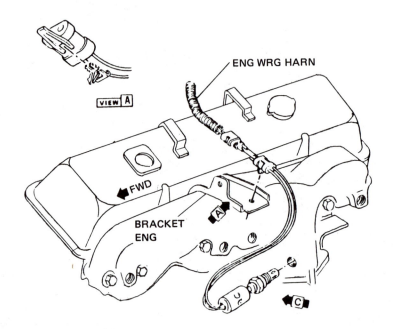

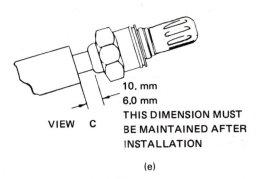

VIEW   C

10. mm

6.0 mm

THIS DIMENSION MUST
BE MAINTAINED AFTER
INSTALLATION

(e)

Figure 13-17   (*continued*)

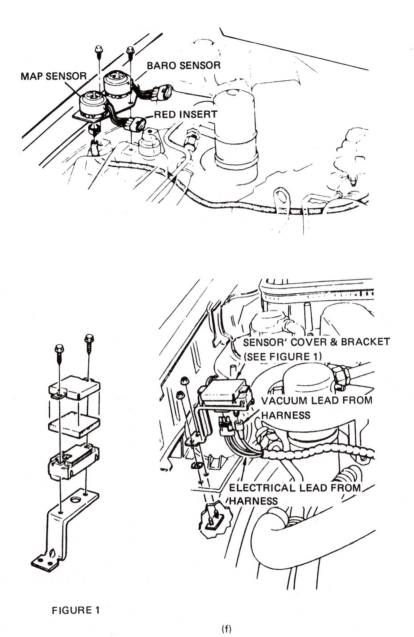

FIGURE 1

(f)

Figure 13-17 (*continued*)

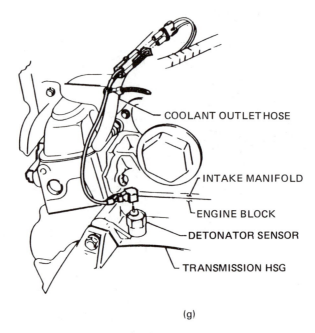

COOLANT OUTLET HOSE

INTAKE MANIFOLD

ENGINE BLOCK

DETONATOR SENSOR

TRANSMISSION HSG

(g)

Figure 13-17 (*continued*)

Ford

1. *Service codes.* As noted earlier, the pattern of thermactor solenoid pulses occurring during the test procedure relates to particular service codes. If the EEC tester is being used, the solenoid patterns will flash on the test display panel. If the tester is not being used, the operation of the solenoids themselves must be observed.

Service code patterns are based on simultaneous pulses of the thermactor solenoids. For instance, the number "1" is represented by this pattern:

both solenoids "on" for ½ second, both "off" for ½ second

The number "2" is represented by this pattern:

both solenoids "on" ½ second, both "off" ½ second
both solenoids "on" ½ second, both "off" ½ second

Both solenoids remain off for 1 full second before the second digit of a two-digit code is signaled. If more than two codes are present, both solenoids remain off for 5 seconds between codes.

Testing usually takes about 1 minute. After the last service code has been signaled, the vehicle remains in self-test operation for another 15 seconds before returning to normal running. Ford service codes are described in Fig. 13-20.

2. *Component locations.* Figure 13-21 shows the locations of major EEC components. As in the GM system, it is very important to make sure that all cables are

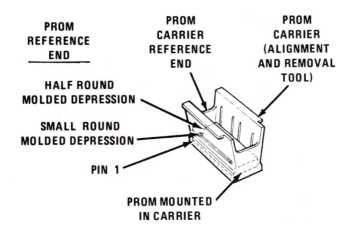

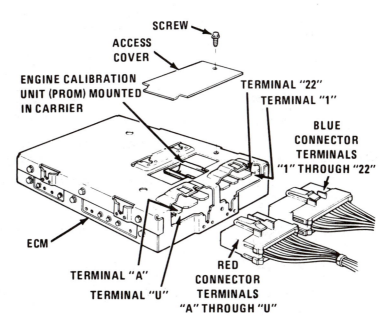

Figure 13-18 PROM installation. (Courtesy of Delco Electronics Division, General Motors Coporation.)

TROUBLE CODE IDENTIFICATION

The "Check Engine" light will only be "on" under the conditions listed below while a malfunction exists. If the malfunction clears, the light will go out and a code will set, except for one condition, that is code 12. If the light comes "on" intermittently, but no code is stored, see this symptom under "driver complaint."

The trouble codes indicate problems as follows:

TROUBLE CODE 12   No reference signal to the ECM. This code will only be present while a fault is present. It will not be stored with an intermittent problem.

TROUBLE CODE 13   Oxygen sensor circuit. The engine has to run for about 5 minutes at part throttle before this code will show.

TROUBLE CODE 14   Shorted coolant sensor circuit. The engine has to run two minutes before this code will show.

TROUBLE CODE 15   Open coolant sensor circuit. The engine has to operate for about five minutes before this code will show.

TROUBLE CODE 21   Throttle position sensor or WOT switch (when used) After 10 seconds and below 800 RPM.

TROUBLE CODE 23   Open or grounded Carburetor M/C solenoid.

TROUBLE CODE 32   Barometric pressure sensor (BARO) output low.

TROUBLE CODE 32 & 55 (At Same Time)   Grounded +8V, V REF. or faulty ECM.

TROUBLE CODE 34   Manifold Absolute Pressure (MAP), sensor output high. After 10 seconds and below 800 RPM.

TROUBLE CODE 44   Lean oxygen sensor. The engine has to run for about 5 minutes in closed loop and part throttle at road load before this code will show.

TROUBLE CODE 44 & 55 (At Same Time)   Open tan wire to oxygen sensor or faulty oxygen sensor.

TROUBLE CODE 45   Rich oxygen sensor. The engine has to run for about 5 minutes in closed loop and part throttle at road load before this code will show.

TROUBLE CODE 51   Faulty calibration unit (PROM) or installation.

TROUBLE CODE 52 & 53   Faulty ECM.

TROUBLE CODE 54   Faulty M/C solenoid and/or ECM.

TROUBLE CODE 55   Faulty oxygen sensor, open MAP sensor, open in sensor return wire to ECM, or faulty ECM.

Figure 13-19   GM trouble codes. (Courtesy of Delco Electronics Division, General Motors Corporation.)

Service code chart

Service code number	Explanation of code
none	No service code output
any	Service code output on one solenoid only
11	EEC system okay
12	Engine RPM is out of specifications
21	Engine coolant temperature sensor fault
22	Manifold absolute pressure sensor fault
23	Throttle position sensor fault
31	EGR Position sensor fails to move open
32	EGR Position sensor fails to go closed
41	Fuel control "lean"
42	Fuel control "rich"
43	Engine temperature reading below 120°F
44	Thermactor air system fault

Figure 13-20 Ford service codes. (Courtesy of Parts and Service Division, Ford Motor Company.)

Figure 13-21 Ford electrical harness and connectors and the location of the major components: EEC system installation. (Courtesy of Parts and Service Division, Ford Motor Company.)

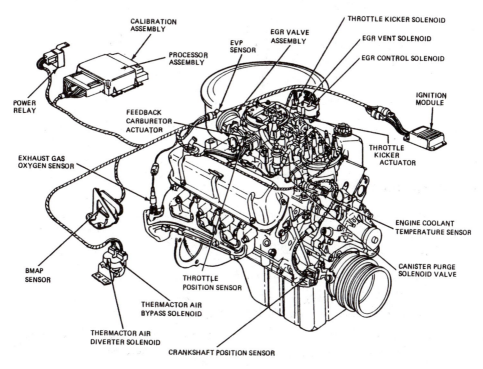

properly and securely connected. According to Ford, very critical connectors should be protected with a grease moisture barrier.

    3. *Connecting the equipment.* Figure 13-22 shows the procedure for connecting the tester to the vehicle.

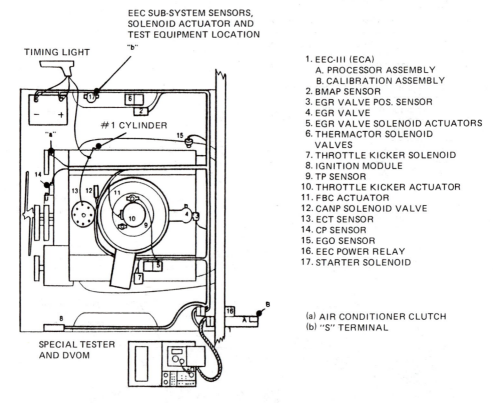

EEC SUB-SYSTEM SENSORS, SOLENOID ACTUATOR AND TEST EQUIPMENT LOCATION

TIMING LIGHT

"b"

\# 1 CYLINDER

"a"

SPECIAL TESTER AND DVOM

1. EEC-III (ECA)
   A. PROCESSOR ASSEMBLY
   B. CALIBRATION ASSEMBLY
2. BMAP SENSOR
3. EGR VALVE POS. SENSOR
4. EGR VALVE
5. EGR VALVE SOLENOID ACTUATORS
6. THERMACTOR SOLENOID VALVES
7. THROTTLE KICKER SOLENOID
8. IGNITION MODULE
9. TP SENSOR
10. THROTTLE KICKER ACTUATOR
11. FBC ACTUATOR
12. CANP SOLENOID VALVE
13. ECT SENSOR
14. CP SENSOR
15. EGO SENSOR
16. EEC POWER RELAY
17. STARTER SOLENOID

(a) AIR CONDITIONER CLUTCH
(b) "S" TERMINAL

Figure 13-22   Test equipment look-up. (Courtesy of Parts and Service Division, Ford Motor Company.)

SAFETY PRECAUTIONS

The same safety precautions that you follow when performing any service operation should be observed when conducting the diagnostic procedures referred to in the preceding test. At the minimum, you should:

1. Block the vehicle's front wheels.
2. Apply the parking brake.
3. Put the transmission selector in PARK.
4. Unless otherwise instructed, turn the acessories off and close the door.
5. Make sure that the work area is adequately ventilated.

Index